The New Farmer's Almanac

T0091687

ADJUSTMENTS AND ACCOMMODATIONS

VOL. VI

GREENHORNS

The New Farmer's Almanac, Vol. VI
Adjustments and Accommodations

Published by the Greenhorns
greenhorns.org

GREENHORNS

Printed in the United States by McNaughton
and Gunn. ISBN 978-0-9863205-4-5

The publication of this book has been made
possible in part by support from Jay Brown.

Editor in Chief
Severine von Tscharner Fleming

Commissioning and Outreach Editor
Renée Rhodes

Lead Editor
Alli Maloney

Poetry and Copy Editor
Jamie Hunyor

Art Director
Amy Franceschini

Lead Designers
Tatiana Gómez and José Menéndez,
Counterform Studio

Chapter Openers Artist
Andrew Long

Proofreaders
Izzy Monroe
Acadia Tucker

Cover Artwork
Current, 2019, Cyanotype print by Melody Joy
Overstreet and Vincent Waring
Wind Theater, 2019, wind-powered typeface
by Futurefarmers

Typefaces Used
Tiempos Text, Tiempos Headline,
and National by Klim Type Foundry

Printer
McNaughton and Gunn

Paper Used
55# Natural Offset Antique, FSC Certified,
360 PPI

DEDICATION

This Almanac is dedicated to our human and nonhuman kin,
those who are moved to migrate as they make adjustments
and seek accommodations in our ever-changing ecosystems.

Fallen Fruit *Fruits from Garden and Field Detail* (image courtesy of the Victoria and Albert Museum)

Z. KNIH ALOISE KUČÍKA

THIS ALMANAC BELONGS TO

_____

TABLE OF CONTENTS

January Colder Than Ever
Rematriation, restoration, stories from Maine

February Causeway
Bridges over troubled water, radical climate adaptations, waiting for water

March Nobody to Serve Coffee
Surplus and bounty, shifts, changing energetic priorities

November   Acknowledging Violence

Hunger, work-life (im)balance, culling, killing, confronting ease

December   Repose

Recovery, rest, facing death, eco-anxiety

→ **Unknown photographer**   *Mississippi Flood* (image courtesy of Library of Congress)

The Almanac

ALMENAK ALMANACH

AL-MANĀKH ALMANAQUE ALMANACK

Renée Rhodes

COMMISSIONING EDITOR'S NOTE

When a very dry fall transitioned into the winter of 2022, large atmospheric rivers roiled through my city and others on the West Coast of the United States and British Columbia, cutting off routes of connection between rural, urban, and transitional places. The farm fields became flooded—too flooded to farm—and the precarious routes that link rural and urban ecosystems and our networked food systems were impassable.

As a Californian, drought and wildfire nerves live in my body. In my West Coast bioregion that springtime, the land dried out, received late spring rain or snowstorms, and then dried out again—over and over, an undecided pendulum. Spring shifted into summer. In early July, I visited a prairie in southern Oregon. At four thousand feet, this upland grassland still had green grasses and wildflowers in bloom, but the land and creek were dried out nonetheless. A deeply channelized creek, the absence of beaver, and the subsequent loss of transitional floodplains to percolate water through the meadow means that the land cannot hold onto water for long. The emblematic rush of energy and opaque blue featured on the cover of *Vol. VI* conveys that feeling, the reception of much needed water but in an overpowering fashion—one that breaks apart our systems and our soil.

In this book, summer moves us through and beyond the boom/bust cycles of a typical harvest season and the grasping at straws for water until the well is all dried up. The writings expand the view of how we define "farming" in the first place and speak to intimate connectivity and grounded relationality with the land in pursuit of personal wholeness. Many authors and artists within this volume explore the human connection to waterways and watery places, reimagining infrastructural possibilities: acequias in New Mexico, oceanic floating farms, and the potential of technological restoration for rivers turned toxic by mining. Philosophical exploration of pruning as an act of practical tending meets science fiction futures; stories of Indigenous Hawaiian farming practices; and commitments to farming with creativity and

intimacy of scale are all within.

I work with others on a small grassland restoration project, and last year, we planted in July against popular advice to plant in fall. We did that because we can no longer rely on the fall rains to come but we can rely on August fog. Along my coastal edge during the summer we are shrouded in thick, wet, and cold fogs. With this in mind, we planted under the dripline of a patch of old invasive eucalyptus trees, whose long leaves captured fog and rained it down on our new grasses all summer long. During the dryness of fall our plants were already well established.

Constant presencing, attention, and an aptitude for perpetual adjustment: these are very useful skills and ones that regenerative farmers and land workers practice everyday. Those who are attuned to the baseline, who are nimble in response, humble in their pace, and calm in a present of constant change and uncertainty are the skilled workers we need in this moment. Among the writings in this book we find flourishing autonomous expressions of gender, sexuality, community, economy, and temporality as they intentionally intersect with land practices that seek to produce conditions of abundance. These lifeways propagate despite the violence embedded in the land and the bodies of so many people, through cultural norms of nonconsent, dispossession, and white heteronormativity.

This volume is not filled with answer keys or bold new visions for the land; instead the adjustments and accommodations modeled come through subtle mindset shifts, attunement to the iterative potential of place-based intimacy, mutual aid exchanged between species, and relationality with the land.

I see other creative land workers practicing their own systems of logic, ones that purge purity drives and perfect plans; that practice a watery ethos: slowing, settling, sinking into the land with the common good sense of love and responsibility; and ones that work to create healthy systems and soil that cannot be so easily broken. ○

Alli Maloney   *Along the Ohio, Adena/Hopewell/Shawandasse Tula (Shawanwaki/Shawnee) Land*

Alli Maloney

MANAGING EDITOR'S NOTE

Where I live, it floods. The creek that splits my driveway muddies and lifts the stones and silt from their positions, higher, higher, spreading them thin. Country roads lower and the muddy water rises, eating the pavement, setting sag to house frames. In late summer cars wait on both sides of an impressive rush until one breaks in, bisects. We move in shifts. *Pacing.* Our butch tires are a privilege. We cheer on the other side, having survived again.

Where I am from in North Carolina, soil is clay. I return to clay now, north, in the Appalachian foothills. Here, bobcat ears at attention in the morning bring mine to the same stance. A fox with the slink of a thigh-high sock moves across pavement into the hollow where we both choose to sleep. Starlings are held by trees and tattoo the sky, extending down to us sensory overload. At ground-level, bugs click, peep, or stay mute.

At the woodline of a forest long-tended by my generous neighbors, I see myself. Infantile garlic mustard presses through the forest floor into warmth. We are all held by the same ground as before; I came for its ancestors across a wide swath of this habitat when I began stewarding a year prior.

Editing this compendium, considering the natural world from behind a computer screen for months, my eyes fixed and unfixed on pear trees in the distance. My legs crossed and uncrossed. Straddling two worlds, on and offline, laboring, I studied.

I became a farmer. I wept in the orchard. The smell of freshly-cut winter barley consumed me. I planted perennials, buried ten raccoons, broke three mowers, rubbed *Artemisia annua* on my wrists, thanked and damned the sky, milled trees, and washed soiled palms of compost they turned to make. I burned my skin in the sun. Native wildflowers changed guard. Next to nigella, I listened; in the tomatoes, a hummingbird suspended in air. But on many days I lay inside, on the floor, terrified for and of the earth.

One among a generation of greenhorns, I am thirty-three and compelled. The elders aged before I knew myself well enough to know that I need them, and now I play catch-up. Untethered I've rushed from coast to coast, job to job, never in a space long enough to consider a season's particulars; I whittled down my daydreams, put myself in skirts and cities, and made a careerist decade out of *just-one-more-year!*'ing it. When I did visit farms or intentional communities, I learned—*parsley, not carrot, but good guess; try hoeing that instead; just cut at the base and leave the roots, they'll break down into the soil*—but assumed I would lose skills as they rusted without use over time, unless these futures, seemingly unattainable, became mine. I wanted to stay every time.

There is benefit to laying no roots, a wide swath of loved ones from each short-term home and what I gained from each. What is lost is the strength found in locking arms and staying put, year after year, amid bounty or ruin. I am not alone in longing for consistency of place and effort. Thirsty with intention, those like me dream of a structure that serves our values but live straddled in liminal spaces. We square up against alternatives that include suffocation: in the dread of the inherited disaster state, amid the proliferating demise of housing and healthcare systems, inside the center of worry about caretaking to come.

In the collapse, though, I see us pledge to create new knowledge through shared values, live friendship as resistance and revolution, and champion radical interspecies connection. Institutional failures can result in interpersonal success. Focus can be re-seized. Despair turns to progress when we antagonize the circumstance we've been born into through conscientiousness, and reimagine, reframe, and reimpose goals. When we see the tree as an opportunity to see one another we thrive underneath its canopy, necks strained in chorus.

Commitment predicates on flexibility and accepting what functions without arms to guide. To commit to landwork is to commit to the plot in its size, is to commit to its potential and its failures, is to commit to the acre as it expands into many and beyond arrogant notions of ownership that allow us to reject responsibility to care for those beyond the borders we draw around ourselves.

The individuals I know newly working for and on themselves include a forester-in-training, curious metalworker, traditional broom maker, passionate self-taught mechanic, construction cowboy, urban flower farmer, chestnut aficionado, and scrappy saver learning to timber frame. Farm punks are committing to habitat. We cross digital divides to support one another until we share the same land. And as I glean knowledge from the region carefully folding me into its womb, I recognize a bounty that has been with me all along. I have known it in my dad's five-gallon-bucket gardens and

enthusiasm for life, which extends from his to my fingers and into the ground below. In morning glory wrapped around sunflowers that returned to the garden at no hand of mine, I found my mother, whose body carried roses to the grave. Assurance has always been part of my path. Revelations, however, come in windfall toward the earth around me, where I rake and shovel and scoop them up with strong hands.

We stragglers are trying, and ready to join you. With these few years of experience and access, I now can extend it to others. Sweeping potting soil from the porch slats, I sing out to my co-conspirators a familiar refrain and let the echo ring: *what's mine is yours.* What tamps the seed is not pressure alone but the possibility of another world. ○

Alli Maloney *Stoned with La Doña at Blue Mountain Center, Haudenosaunee/Kanien 'kehá: ka Land*

Colin Sullivan-Stevens *Grey Lodge*

Severine von Tscharner Fleming

WELCOME

For fifteen years I have helmed this little ship called Greenhorns, now safely moored on the Pennamaquan River, a flat-bottomed estuary in the heart of Cobscook Bay, in Washington County, Maine. This little ship has carried many on our journey, down the Hudson River; around the Puget Sound; between the California Redwoods in Granges; across the Upper Midwest with nuns and anarchists; across the dusty irrigated deserts and up into the basins and ranges with green cowboys, down to the gritty cities of the Atlantic South; through museums, barns, basements, drafty conference centers, and lakeside cherry orchards. Everywhere, we have found bold, thoughtful collaborators and farmers. On behalf of the Greenhorns project, thank you all for your extraordinary spirit and contribution to the twinkling campfire of this project.

Greenhorns was founded and exists to enliven the cultural world of young farmers. Our films, publishing, parties, podcasts, and exhibits are all about recruiting, cherishing and stimulating young humans who gravitate to the land. We celebrate those who sink their fingers and thumbs and lives into loving work, heaving up bright orange carrots, collecting plums, slicing cabbages, washing the dishes, packing the truck, popping up the tables, appreciating the dew tips and parsnips and Jersey hips—enough to stick with it.

Do such actions and feelings express a premonition for a world that is changing? A longing? A coded vibration for human survival that quivers our timbers? Certainly, it is not just romance, and no one can take lightly that it doesn't pay what it should. We can say with confidence that the will to farm, heal, produce, and pollinate must be a kind of powerful programming if it can persist alongside, underneath, or despite the Disney and Barbie-marketing onslaught of American childhoods, heavy metals contaminating our food and soil, and the sensory deprivation of indoor education. And persist it does. Sometimes more and sometimes less, but every year more young farmers join the ranks.

A hefty tranche of folks over sixty keep the age of the average farmer high, but the joiners keep joining, and much of what is needed is help transitioning between generations and operations. I wish there was a whole section of the Farm Bill devoted to land transition and facilitation. There will be lots of funds called 'climate adaptation' that are in fact a bail out for operations marooned by drought and flood, but perhaps some portion of it can be siphoned out to help pay for the infrastructure, diversification, and re-regionalization that will in fact yield us a more sustainable food system. It is unlikely that we get through this period without shuddering fallout and crisis in our food system.

Fundamentally, it is the belief of our organization that without a massive infusion of human ambition into the practice, reform, and transformation of agriculture, the human project is doomed to crash our planetary ecosystem. In an economy rotten with hype and speculation, corrupted by lobbyists and subsidies, riddled with the scabby tick bites of extraction and violent fantasia of free-loading plastic hyper-suck capitalism, it feels good to kibbutz with others who have tuned their humanity, lives, and tempo inwards to focus on horticulture, gentle and committed stewardship, cherishing the materiality of fruit sugars, as well as thought, fermentation, and ambition.

The premise of Greenhorns is that this calling to farm is best felt in good company, where many souls can feel the feeling together. That we'd better make a scene and a splash, some good soup, hot gossip, jazzy posters, dioramas of fungi and sheep's wool, taxidermied goose wings with LEDs behind them, a reading area and flower bouquets to celebrate the entering class. The book you hold in your hands, is itself a field's edge flower bouquet with artists, bright green freshman, old hands like rose hips with fruit attached, retired professors keeping an oar in each, navigating with fresh curiosity the drama and pathos of agrarian life amid of the sixth great extinction. All these voices piled up to give company to the others, to the readers, to the young people.

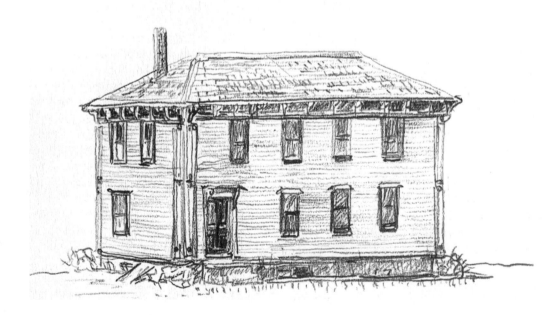

Colin Sullivan-Stevens *Reversing Hall*

Only together can we actually pull it off. It may come across as judgemental to those of our readers whose experience of what we're talking about remains an abstraction—and I mean it provocatively. What is required is not just shopping local, but a complete transformation of our human culture and orientation. It means committing to participate with the land and the living world. Moving alongside and through our differences of identity, affiliation, opinions, and projections, we learn to hold up our little parts of the world, hauling wheelbarrows to recover the last of the tomatillos in the sunshine, feeding the bees to last through the winter, protecting the fruit trees from winter mice, doling out the compost. It is about the humble application of effort towards diversity, productivity, elegance, inter-relation. Drinking water from the ground, eating food from nearby, living a smaller and more beautiful life. If more of us do not figure out how to bend ourselves into the kind of human who can live with a more modest percentage of the world's energy, metal, and Amazon-delivered amenities, then we probably deserve our fate. It is hard to imagine a moral superiority that entitles us to destroy the world.

And what is the opposite of that? Binding ourselves, our bodies, imaginations, and efforts to the destiny of our home river, our hometown. In becoming "at home" with the miracle of life that surrounds our little Mammalian circulatory systems of consciousness and the circuitry of our loving relations, we will get nourishment from the plants and the air.

To succeed, our movement for food sovereignty, agroecology, and regenerative engagement with the living world needs a steady and increasing stream of hopeful, hardworking committed humans. We also need the cultural and political valuing of those human efforts. It is my hope that this shockwave of COVID-19 that ripped through our bodies and economy gave each of us a change to tune into the void within ourselves, to the humility of our physical power, to, in some way, release from the projections and propaganda and flim-flam of the mediated world. I hope we begin to listen more closely for the signals within our home ecosystem, to the sensors in our brains far older than language or reason, to accommodate ourselves to the changes we will or are experiencing, to adjust our expectations, and to embed ourselves wholeheartedly inside the enveloping strength of our community. To commit to our love of the world even in its unraveling.

Homecoming for Greenhorns

For the past five years, the impulse to make home has been strong within me and for Greenhorns. Our campus is basic and rustic in parts but the elements are coming together. Within a tiny, historic, super-rural coastal town, along the river where the town meets the sea, we have built many of the structures and infrastructures needed for land-based cooperative living—the kind of living we think is most feasible in the coming age. Having been a party organizer and a media organization, it's quite a change for me to put on a contractor hat as a greenhorn historic preservationist, adaptive re-user, stripper of paint, and purchaser of thousands of feet of greenhouse plastic. But the urge was undeniable, and so here I am, soon to welcome a little baby with husband Terran Welcome, with my fingers crossed about a next generation of leadership within the organization.[1] Already we've got children crayoning in the sunshine, kneeling between the stacks at the library, sitting up to eat strawberries with cream on chairs we have dusted off and repaired, making beach trash puppets, rolling fire-weed between our hands to make Russian tea, pressing apples to cider, gathering in the golden afternoon. This small agrarian universe we've created has taken a life of its own.

Since 2017, we've pulled together a timber frame summer kitchen to gather and cook in and a stately Odd Fellows Hall with velvet benches, thousands of books, and surround-sound for movies. In the old mining office by the river we've created a pine-paneled farm store with locally grown provisions, our own organic produce, and value-added products, frozen meat and fish, delicious cheeses—even bubbly drinks. In the next room, there's a sunny office for Greenhorns; a parlor and guest houses for artists, winter farmers, summer farmers, and guests. Across the street, a commercial kitchen with freezers, grinders, fermentation and dehydration tools, and cupboards full with cherry-pitters and toma-to-squeezers. A picnic area, a little village of kelp-drying greenhouses, a mushroom lab, a lengthening network of trails; ever larger gardens and orchards; our own organic cran-berry bogs, blueberry barrens, long swaths of perennial plantings for bees, and for bouquets. And also, crucially, a collection of little houses along the river: Fox, Elver, Cat, Eagle, Mustard, Snapper—each long empty, each making its way back into liveability.

As well as the home farm, our cultivated area is now approaching two acres with well established garden beds and more than five hundred fruiting trees, shrubs, and bushes. Don't forget the wildly seeding momentum of what likes naturalizing here: carraway, borage, anise, mustards, daikons, arugulas, calendulas, poppies, chives, amaranths, coreopsis, indigo, service berries, Japanese chrysanthemums, and flanks of yarrow. As you drive up the road,

there is sea buckthorn, elderberries, cherries, persimmons, quince, apples, plums, pears, peaches, and walnuts, coming along as fast as we can compost them. Our little agro-forest has begun to bear fruit and in the coming years we will have our hands full with uninterrupted harvest from June through October.

For those of you new to the Greenhorns project, or who only read these Almanacs we produce, and do not watch the films, listen to the podcasts, attend the workshops, and track our newsletters or updates, you may not know about this wonderful campus and the "permanent home by the sea" we have been making for Greenhorns and our participants.[2] Our goal is to have the space and facilities to house all the teammates and visiting learners to grow our own food, fish our own fish, produce our own media and books, to offer classes to locals and visitors, as well as to welcome more young farmers to this region. We want to help more locally born youth get involved with natural resource conservation and livelihoods, in organic agriculture, and entrepreneurship, thereby improving the resilience, contentment and quality of life in this hamlet.

Greenhorns was founded in 2008, then based in a half-dozen rented offices and sheds. Our activities are now spread out across an amazing array of buildings and parcels of agricultural and forest land operated by Smithereen Farm. If you came in the begin-ning, you will be amazed at the midway-to-great we have arrived at in year five. Every year we welcome more artists, teachers, campers, and farmers. The opportunity to gather together such a collection of houses, meadows, and gathering places is a rare gift for a small and young organization like ours, a more common thing to see in the 1970s! It has been made possible because of our remote location, the financial support of a close relative, amazing humans arriving to contribute their talents, and the pent-up drive of ten years of farmlessness.

The coming months and years will present more roofs to be fixed and foundations to be stabilized; a solar array and electric car charging station, much-needed paint jobs on many of our historic buildings.[3] The work continues but we are within sight. We look forward to welcoming more visitors and leadership into cozy and freshly painted homes and a vibrant village of books, bikes, deep garden beds, strong moorings, sufficient hand-washing sinks and drying racks—the means to make a good life together, whatever the weather holds in store. Those of you who helped with the startup-phase of this new campus: thank you.

To watch a film about Greenhorns at fifteen, including a campus tour, visit the capital campaign page at: *greenhorns.org/15years* or watch the films and learning journeys about fish, berries, and natural history curriculum we started making during COVID-19 at *earthlife.tv*. Or read the position paper about how to protect our seaweed commons from industrialization at *seaweedcommons.org*. The programmatic engines of Greenhorns continue to buzz away at full speed, summer workshops with returning alumni teachers in wild foods, seaweeds, insect habitat, and timber framing. Thanks to a major USDA grant we can coordinate more than before.

We are here, at the top right corner of the United States as it juts out way far east after a zigzag of bays, inlets, islands, and massive peninsulas—Mt. Desert, Great Wass,

Gavin Zeitz   *Map of Smithereen Farm*

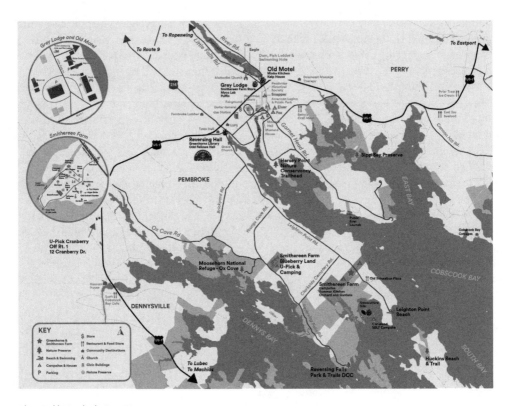

Ginny Maki *Pembroke Town Map*

Roque Island, Grand Manan—where the cold waters of the Labrador Current cruise down, and the warm waters of the jet stream rush up, mixing and pulsing with planktonic power that constitutes one of the more powerful marine contexts on the planet. We are located on an important swimway and flyway, with migrant avians from the arctic and the tropics dropping in to gorge themselves on berries, herring, sand fleas, or thistle seeds, pausing to fluff their feathers in our patches of dust and rear their baby bobolinks in our front meadows. As with other major coastlines, it forms a pivot point and magnetic line of orientation. The twirling hurricanes make the news, but don't forget the birds and whales cruise along the upwelling of food and energy, too.

The productivity of sunshine makes life on this planet and in concert with ocean nutrients, the sun shines to make this place one of the richest fishing grounds on Earth. It is the plankton, microalgae, and macroalgae whose bodies make the majority of oxygen in our atmosphere, whose buffering and carbonaceous micro-skeletons protect us humans from an even more rapid spiral of climate consequence. It is these last-functioning ecosystems, full of living, cycling, carbon-based native fisheries, bird flocks, whale pods, and all the associated forest and earthlife that offer us our best hope for survival. The Gulf of Maine is such a place of life. What a privilege to make this the Greenhorns' home.

We are also in the county with the highest rate of youth departure in the state of Maine. We have high biodiversity, the presence of native fish stocks in most of our rivers, hard-working and resilient neighbors, but also some depressing statistics on drugs, crime, child poverty, and literacy rates. When you visit you will see: the community is strongly bonded through school and bean suppers, but there remains much to do. We need a nursery school, a native plant nursery, a tree nursery. We need more civic interaction and intergenerational attention on planning for local resilience, planning to protect us from mounting development pressure, and thinking about how solar developments around rivers might be best sited. The many empty, abandoned, and derelict houses in our town seem strange in a world so full of homeless people. With such a crush of human density and intensity, it is confusing that this far-flung corner of the world has so much vacancy. Of course, there are so many reasons, economic opportunity chief among them. The prospect of more tourism and more retirees does not satisfy. It feels clear that a big part of our work here is to make more opportunities for the kids who gather here, to pick strawberries and blueberries, or to take their date for a kiss on a bench by the river, screened by a wall of willows while the eagles whistle together above the treeline.

Operating in anticipation of changes, the needs of the land, and our lives upon it, we adjust our expectations and alter our approach-es. We at Greenhorns, and many of the writers in this Almanac, recognize that the world has been much altered in the past few years of COVID-19 and the intensification of injustice it has brought. We recognize that we can alter course and help others alter course as well. We recognize that the greater our capacity for imagining this anticipatable future, the greater our power to shape it is.

At Greenhorns, we look forward to involving many others in that anticipation, in the spirit of a shared understanding about how much we need each other and about how good the work feels to do. ○

Notes

1. Already we are blessed with many longtime team mates: Alex Plowden in the office, Renee Rhodes at the Almanac wheel with Alli Maloney, Elena Bird in remote capacity steady at her oars, the wonderful Charlotte Watson returning for another season. John Stevens and Nathalie Jeremijenko holding the engineering department together—when you greet them, please thank them for all their efforts at maintaining, repairing, and constructing the ecosystem of this campus, making us water-proof, wifi-connected and able to host you.
2. @thegreenhorns @smithereenfarm @reversinghall @maineseaweed
3. Being strongly rooted in a home-place brings with it responsibility for "off-topic activism." This past year we found ourselves forced to get involved with a new issue: fighting a Canadian polymetallic mine that wanted to open in our watershed. We learned what to do, hired a wonderful lawyer from Drummond and Woodsum, built up coalitions and organized teach-ins, and poured thousands of hours into this work. Finally, in May 2022, our town voted to pass an ordinance that prohibits industrial scale metallic mining. You can learn more at *pembrokecleanwater.com*.

Alli Maloney *Smithereen Farm, Passamaquoddy Land*

2023

Celestial Calendar

Calendars and Data

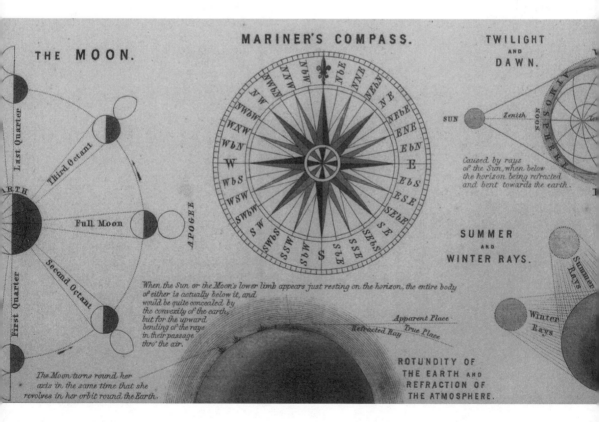

Adam and Charles Black, Sidney Hall, and William Hughes *The Solar System and Theory of the Seasons* (excerpt) from *General Atlas Of The World: Containing Upwards Of Seventy Maps*

PHASES OF THE MOON

	Sunday	Monday	Tuesday	Wednesday	Thursday	Friday	Saturday	Sunday	Monday	Tuesday	Wednesday	Thursday	Friday	Saturday	Sunday	Monday	Tuesday	Wednesday
JAN	1	2	3	4	5	6	7	8	9	10	11	12	13	14	15	16	17	18
FEB			1	2	3	4	5	6	7	8	9	10	11	12	13	14	15	
MAR			1	2	3	4	5	6	7	8	9	10	11	12	13	14	15	
APR						1	2	3	4	5	6	7	8	9	10	11	12	
MAY		1	2	3	4	5	6	7	8	9	10	11	12	13	14	15	16	17
JUN				1	2	3	4	5	6	7	8	9	10	11	12	13	14	
JUL						1	2	3	4	5	6	7	8	9	10	11	12	
AUG			1	2	3	4	5	6	7	8	9	10	11	12	13	14	15	16
SEP						1	2	3	4	5	6	7	8	9	10	11	12	13
OCT	1	2	3	4	5	6	7	8	9	10	11	12	13	14	15	16	17	18
NOV			1	2	3	4	5	6	7	8	9	10	11	12	13	14	15	
DEC						1	2	3	4	5	6	7	8	9	10	11	12	13

Tatiana Gómez *Phases of the Moon*

Thursday	Friday	Saturday	Sunday	Monday	Tuesday	Wednesday	Thursday	Friday	Saturday	Sunday	Monday	Tuesday	Wednesday	Thursday	Friday	Saturday	Sunday	Monday
19	20	●	22	23	24	25	26	27	●	29	30	31						
16	17	18	19	●	21	22	23	24	25	26	●	28						
16	17	18	19	20	●	22	23	24	25	26	27	●	29	30	31			
●	14	15	16	17	18	19	●	21	22	23	24	25	26	●	28	29	30	
18	●	20	21	22	23	24	25	26	●	28	29	30	31					
15	16	17	●	19	20	21	22	23	24	25	●	27	28	29	30			
13	14	15	16	●	18	19	20	21	22	23	24	●	26	27	28	29	30	31
17	18	19	20	21	22	23	●	25	26	27	28	29	●	31				
●	15	16	17	18	19	20	21	●	23	24	25	26	27	28	●	30		
19	20	●	22	23	24	25	26	27	●	29	30	31						
16	17	18	19	●	21	22	23	24	25	26	●	28	29	30				
14	15	16	17	18	●	20	21	22	23	24	25	●	27	28	29	30	31	

PRINCIPAL PHENOMENA 2023

Phenomenon	Date	Time
Perihelion	January 4	08:17 PST
Equinox	March 20	14:24 PST
Solstice	June 21	07:58 PST
Aphelion	July 6	13:06 PST
Equinox	September 22	21:50 PST
Solstice	December 21	19:27 PST

Data drawn from Sea and Sky's Astronomy Reference Guide, US Naval Observatory, *The Old Farmer's Almanac*, the American Meteor Society, and Archeoastronomy

January 3, 4 Quadrantids Meteor Shower

April 22, 23 Lyrids Meteor Shower

May 6, 7 Eta Aquarids Meteor Shower

May 5 Penumbral Lunar Eclipse
A penumbral lunar eclipse occurs when the Moon passes through the Earth's partial shadow, or penumbra. The Moon will darken partially but not in full. The eclipse will be visible throughout all of Asia and Australia and parts of eastern Europe and eastern Africa.

July 3 Full Moon, Supermoon
The Moon will be located on the opposite side of the Earth while the Sun and its face will be fully illuminated. This full moon has been known as the Buck Moon, Thunder Moon, and the Hay Moon. This marks the first of four supermoons for 2023. The Moon will be near its closest approach to the Earth and may look slightly larger and brighter than usual.

July 28, 29 Delta Aquarids Meteor Shower

August 1 Full Moon, Supermoon
The Moon will be located opposite the Earth from the Sun and will be fully illuminated as seen from Earth. This full moon has been known as the Sturgeon Moon, Green Corn Moon, and the Grain Moon. This is also the second of four supermoons for 2023.

August 12, 13 Perseids Meteor Shower
The crescent moon should ensure that skies are still dark enough for good viewing. Best visibility will be from a dark location after midnight. Meteors can appear anywhere in the sky, especially radiating from the constellation Perseus.

August 31 Full Moon, Supermoon, Blue Moon
The Moon will be located on the opposite side of the Earth as the Sun and its face will be fully illuminated. This is the third of four supermoons for 2023. As the second full moon in the same month, it is sometimes called a Blue Moon.

September 29 Full Moon, Supermoon, Harvest Moon

The Moon will be located on the opposite side of the Earth as the Sun and its face will be fully illuminated. This Harvest Moon is the full moon that occurs closest to the September equinox each year. This is the last of four supermoons for 2023.

October 7 Draconids Meteor Shower

October 14 Annular Solar Eclipse

An annular solar eclipse occurs when the Moon is too far away from the Earth to completely cover the Sun. This results in a ring of light around the darkened Moon. The Sun's corona is not visible during an annular eclipse. The eclipse path will begin in the Pacific Ocean off the coast of southern Canada and move across the southwestern United States and Central America, Colombia, and Brazil. A partial eclipse will be visible throughout much of North and South America.

October 21, 22 Orionids Meteor Shower

November 4, 5 Taurids Meteor Shower

November 17, 18 Leonids Meteor Shower

December 13, 14 Geminids Meteor Shower

December 21, 22 Ursids Meteor Shower

Adam and Charles Black, Sidney Hall, and William Hughes
The Solar System and Theory of the Seasons

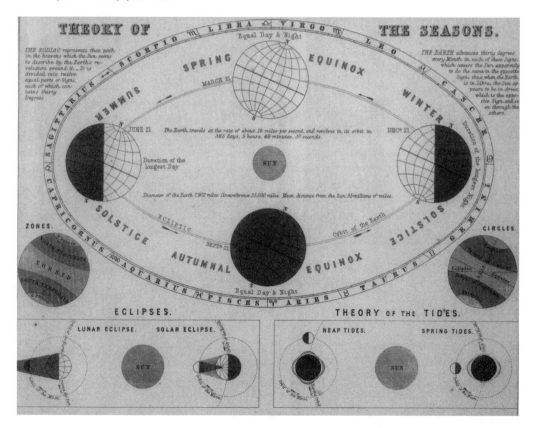

JANUARY

Colder
Than Ever

Rematriation,

restoration,

stories from Maine

Alivia Moore

REMATRIATION

The Future with Indigenous Land

Only take what you need.
—Gluskape

The future of the land is its liberation. The earth will always take its course to restore balance, and there is an opportunity for us to collectively embrace this reality. We can choose to mirror the earth's propensity for balance, and, with that, we can allow for the possibility of human liberation. What we must do is respect the land.

We must humble ourselves to understand that we can't heal it. We are not the earth's saviors. We may make claims to the earth by imposing upon it our laws and our private and collective ownership structures, but ultimately the land can not be owned. We are derived from the land; it holds us rather than any of us truly being landholders.

In Wabanakiyik—so-called Maine and beyond—Wabanaki are rematriating: restoring our matriarchies and embracing the correlating laws and relationships of this land again. As a member of the Penobscot Nation, I see Wabanaki social, economic, and political systems as reflections of a deep relationship with the earth and waters of this place, of humbling ourselves to its cycles.

Prior to colonization, our societies were more balanced, egalitarian, resilient, and responsive. Wabanaki worldview is one of understanding all things in cycles—whether it be seasons, life/death, or village locations changing with abundance of foods. Our cultures reflect the earth, respect natural limits, and are simultaneously predictable and flexible. Wabanaki languages demonstrate our expansive non-binary cosmology. Rather than gendering beings with concepts of "he" and "she", we generally use "they," which means something is alive. We know that all beings— human, plant, rock, animal, and so on—have a gift and a role to play in service of others.

In matriarchies decision-making is based on nurturing and honoring everyone's gifts. Power is derived from a different source than it is in today's dominant culture, where influence over others and resources is paramount. Matriarchies recognize power from within and

Alivia Moore
My daughter harvesting cranberries

June Sapie   *Medicines in the birchbark bowl*

cultivate reciprocity that strengthens the collective. These systems of human law lay out our responsibilities to the earth and one another, rather than our rights and entitlements.

Indigenous people know that food and healing systems are avenues for transforming society. Maine farming and fishing communities must wholly reckon with the past because

it shapes our current reality. It is long overdue that power relationships to, and on, the earth are examined. The way we relate to the earth is how we relate to each other.

Climate crises, social inequities, and racialized violence often leave us feeling entrenched in destructive pathways of exploitation, white supremacy, and colonialism. Remember: the

American project is in its infancy. The Wabanaki nations of the Abenaki, Maliseet, Micmac, Passamaquoddy, and Penobscot exist and were developed in reflection of this land for at least twelve millennia. We are the elders here. We carry essential leadership for collective healing.

To attain liberation on the land, it is necessary to grapple with settler colonialism, white supremacy, and to understand agri-culture as a tool of historical and present-day genocide of Wabanaki people. Burgeoning solutions in the agricultural community—regenerative agriculture, permaculture, and community land-holding, for example—are Indigenous practices that were forcibly disrupt-ed by colonization. While these approaches are a shift toward relationally respecting the land, how they are embodied within white-led movements differ from Indigenous practice in crucial ways. Wabanaki "regenerative" food systems are time-tested and developed hyper-locally in reflection of the specific parts of the earth where we have thousands of years of reciprocal relationships.

Indigenous and Black communities are the most immediately impacted by existing extractive relationships to the land; we can more clearly see the tactics of oppression and harm. Together, Indigenous and Black folks must lead land justice work, and white folks can do more to dignify our struggle and align with our movements.

To come into relationship with the land as a relative, colonial private land ownership must end. The rightful return of land to Wabanaki nations and Wabanaki-led land trusts should occur. Wabanaki, Black, and Brown peo-ple must have access to land and water—the spaciousness to breathe and "be" will sustain us as we lead.

Reciprocity and responsibility need to become central to how we exist with the land and how we organize society. Non-Wabanaki can respect that you are visitors here and know that Wabanaki people, in accordance with our responsibility as your host, will make sure your needs are met in our territory. Trust the leadership of Wabanaki peoples; we are here continuing the work of our ancestors—this work benefits all that call Wabanakiyik home.

Just as we share this land, we share a future. The earth, its land and waters, is the basis of the survival of all beings. The earth is our only source of true security. Land is also at the heart of extraction, imperialism, and environmental collapse. We invite our non-Indigenous relatives into the process of unsettling: dismantle the harmful ways of relating and position yourselves to adapt, as Wabanaki people have done since time immemorial. ○

Sharifa Oppenheimer

AIR

We grow translucent
with imprint of wind.
Conjoined Spirit
  composed of god's breath and
  signatures of animal kin
opens our gills-become-lungs.
Oxygen cycles
through geologic time
through an elongated
history of evolving bodies'
animate tissue.
Breath is a tender sharing
honed through millennia
  coevolved with conifers
  tree frogs, great blue heron
  black bear mothers
  wind rushing down canyons.

Vapor, breeze, gust, zephyr:
living air
lived inter-being.
To be a gnat in this
amaranthine wind
  vanished inside the whirling
for a moment
we are one photon
in the eye of god.

A version of this profile with the same title was initially published in A Litany of Wild Graces: Meditations on Sacred Ecology *(Red Elixir Press, 2022), and was printed by permission of the author.* ○

Hilary Irons *White Window*

Brett Ciccotelli

DESTRUCTION/RESTORATION

The concrete base of the abandoned power-house blocked fish from ascending the river. For decades the river was routed through the powerhouse and forced to turn a hydroelectric turbine that generated power for the local energy grid. When that grid connection was made in the 1940s, the connection between Maine's Meddybemps Lake and Cobscook Bay was broken.[1]

That link remained broken for nearly seventy-five years. Still the fish returned, their numbers reduced but full of potential, waiting for the river to reopen. Some spawned below the powerhouse. Others worked their way up a small side canal. Lucky ones were caught in nets and dumped upstream while unlucky others went off to bait lobster traps or fill cat food cans.

The powerhouse and its generator were built by a mechanical genius whose brilliance nearly destroyed the river and poisoned the earth along its banks. His descendants put blame aside. They allowed state, federal, tribal, local, and nongovernmental partners to bring the people and tools together to restore the river.[2]

On a hot mid-September day in 2021, the excavator's seventeen thousand pound hydraulic hammer shattered the powerhouse's concrete walls into boulder-sized chunks. As these broke apart they dropped onto the heavy wooden crane mats suspended over the riverbed below. Though the concrete—poured to hold back the river three-quarters of a century before—was solid, it crumbled easily under the diesel-powered hammer's fifty-two thousand pounds of impact pressure. As the powerhouse gave way, automotive antiques emerged from the walls. Chrome bumpers, truck frames, and ball joints had been set in the side of the powerhouse to reinforce the concrete walls. Looking down into the demolition site was like staring into the Grand Canyon of automotive history. These relics were piled up and sent down the road to a scrapyard owned by the grandson of the man who created the powerhouse—a steel savings account passed between the generations.

The land adjacent to the powerhouse was operated as a salvage yard and dumping ground for some of the twentieth century's worst chemicals, including polychlorinated biphenyl (PCB)-laced oil from old military equipment.[3] When excavating the contaminated land—a federally controlled superfund site—a Wabanaki village was unearthed within the poisoned ground. Tools and other signs of habitation were found mixed with the toxic chemicals. The ground alongside the river was rich with the memory of how the Wabanaki people lived along and with this river.[4] The pollution, its harm, and its cleanup wove a dark web that reconnected the Passamaquoddy to the site.[5]

Today, after years of remediation and the injection of PCB-eating bacteria into the earth, the site is nearly clean and nearly safe. As the last few clean up treatments are completed the State of Maine, owners of the property since the cleanup began, has returned the land to the Passamaquoddy Tribe. Large stone steps have been installed into the river and into the bank. These steps allow easy passage for river herring, sea lamprey, and Atlantic salmon migrating upstream into the clear waters of Meddybemps Lake and for people to revisit the river's edge.

When the powerhouse was destroyed, Meddybemps Lake's connection to Cobscook Bay was restored.

Across the world there are rivers that need help, damage that needs repair, and land that needs to be recovered. Someday the excavator engines and hammers will be silent, the bioremediating bacteria will be satiated, and children will play along rivers filled bank to bank with fish shaded by the new green leaves of yellow birches. That time is now, but also not yet. ○

Notes

1. Kenneth F. Beland, James S. Fletcher, and Alfred L. Meister, *The Dennys River: An Atlantic Salmon River Management Report*, (Bangor: State of Maine Atlantic Sea Run Salmon Commission, 1982).

2. Fred Bever, "Passamaquoddy Tribe Re-Acquires Land Whose Former Owner Was Called 'Maine's Most Wanted Polluter,'" *Maine Public Radio*, September 22, 2021.

3. Bever.

4. Donald Soctomah, *A Visit to Our Ancestor's Place: Meddybemps-N'tolonapemk Village*, (Passamaquoddy Tribal Historic Preservation Office, 2005).

5. Edward M. Hathaway to Liyang Chu, memorandum, January 8, 1998, *semspub.epa.gov/work/01/3450.pdf*.

Brett Ciccotelli   *Restored Dennys River at Meddybemps*

Maia Wikler

THE BEATING HEART
OF THE YUKON RIVER

I wish you could see the empty smokehouses and quiet fish camps, and understand.

I wish you could feel the crunch of the forest floor
beneath your feet, scorched by unprecedented heat.

I wish you could smell the sour methane on the banks of the Yukon River from permafrost melt, and stand with me, watching, helpless, as the eroding ridge side collapsed into the river, leaving a billowing cloud of silty dust in its wake.

I wish you could feel how hard it is to breathe through silent, choked tears in an emergency community meeting as an elder stands in the room, crying about the salmon fishing ban, the subsistence ban, the native way of life ban.

Maia Wikler *Salmon Heart*

Maia Wikler
Salmon Drying

I wish you could hear the joy of six-year-old Braeylin gleefully shrieking, "Baasee!" (Thank you) to the white fish in her hands, who gave their life in the four-inch mesh net.

I wish you could taste the depths of generosity and community in the bite of a single smoked salmon strip, gifted by a family who will need to survive the winter off of the four fish they could catch this year.

I wish you could hear the unwavering resolve, the clear certainty of these tribal advocates who know

The only way forward is Indigenous governance, liberation from the shackles of a thankless system that in just sixty years has nearly killed off the beating heart of the Yukon River—the salmon.

Sue Van Hook

ENTER MYCOBUOYS

What Do Mushrooms Have to Do with Derelict Fishing Gear, Plastic Pollution in the Ocean, and Aquaculture in Maine?

I spent most summer days on North Haven Island in Penobscot Bay helping my grandparents, Jim and Louise Van Twisk—Poppy and Weezie. I was fascinated watching Poppy tinker: mending bait bags, building lobster traps, pouring molten lead to form fishing sinkers or carving wooden lobster buoys on his Shopsmith lathe. The shapes of the buoys were unique. Some resembled acorns, others looked like torpedoes. Navy blue, buff, and white paint—and license number 7691—unified them. While I never carved one of these buoys, I often painted them. One of the jobs I had on lobstering days when the whole family went along was to hook the buoy out of the water and pass it along to Poppy.

Fast forward to the 1970s, when closed-cell plastic foam buoys made from Expanded Polystyrene (EPS) began to replace wooden buoys. Foam buoys were less expensive, easier to handle while hauling traps, and floated higher in the water. They required painting less frequently and probably cut fuel costs by not adding much weight when setting or taking out traps at the beginning and end of the season.[1,2] After all, it was the era of "Better Living Through Chemistry," according to Dow Chemical Company's advertisement slogan.[3]

One of the cons was that boats of all sizes, including island ferries, could simply plow through a mess of foam buoys if they were in the channel. The impact of a foam buoy against a boat hull was not a big deal; the buoys bounced off the hulls leaving a trace of colored paint at most. In prior years boat captains would carefully navigate around wooden buoys. The outcome during the past three decades is that many foam buoys are severed from trap lines and wash ashore as derelict gear.

From here, issues regarding plastic pollution in the oceans and shoreline soils arise. Maine law states no one but the owner of the derelict gear can touch it, but visitors to the state love to collect buoys along the shore to adorn their sheds back home as tokens of their trip to vacationland. Not only is this illegal, but it also makes coastal cleanups difficult. Additionally, EPS buoys degrade in sunlight over time and fragment into tiny particles that eventually become nanoparticles in the water column. Filter feeders such as clams, mussels and oysters, as well as fish, were found to contain polystyrene nanoplastics.

So while plastic buoys have made life easier for those who fish, their ecological impacts have not been accounted for until recently. How long do we wait for plastic microfibers to

bioaccumulate in the marine food web? The need for new alternatives to plastic fishing and aquaculture gear is now.

Serving as the Mycologist for Ecovative Design between 2007 and 2016, I learned to listen to what fungi have to teach us about their properties, buoyancy in particular. In 2011, my strong ties to Maine led me to grow the first mushroom buoys to replace plastic buoys.

Additional applications of buoyant properties of fungi include Paul Stamets' Mycobooms using saltwater-adapted strains of the oyster mushroom, *Pleurotus ostreatus*, to corral and remediate ocean oil spills and Ecovative's interlocking MycoFoam™ ring to cushion the National Oceanic and Atmospheric Association's DART monitoring buoys against launch impact.[4,5] The ring was designed to

Alex Plowden *MycoBUOYS workshop at Smithereen Farm*

remain interlocked through deployment of the DART buoys, then break apart and decompose into the marine environment. The degradation took five months. Both of these prior applications were solutions to temporary problems.

I began studying mushrooms in 1974, while attending Humboldt State University in Arcata, California. I returned to Maine as Director of Stewardship and Land Conservation for Maine Coast Heritage Trust. At that time, I observed that one could practically walk across the sea on lobster buoys. In 2015, I spent the summer with the Rozalia Project for a Clean Ocean, removing and documenting plastic trash from remote Maine islands. The next year, we collected water samples the lengths of the Hudson River and Long Island Sound to assay for nanoparticles of plastic pollution.

Abigail Barrows, a marine research scientist and aquaculturist on Deer Isle, Maine, performed an analysis of the surface grab samples from the Hudson River and Long Island Sound that led her to compare them with neuston tow net sampling methods. She published the results in *Analytical Methods* in 2017 to help standardize sampling methodology for assessing microfiber particles in different size classes.[6] Using these two sampling methods, Barrows was part of a team that determined where on the globe microfibers are being concentrated.[7] The researchers estimated that the average density of floating microfibers is 5900 +/- 6800 items per cubic meter in the ocean. Bioaccumulation of these microfiber nanoplastics in the food web are causing oxidative stress, neurotoxicity, and negative effects on the consumption, behavior, reproduction, and survival of marine organisms.[8,9]

Between 2011 and 2016, I tested dozens of agricultural byproducts, including kapok—a fluffy seed pod fiber collected from the fruits of the tropical ceiba tree that was used in the life jacket I wore as a child—as well as corn, kenaf, and hemp. Hemp is a marine-resistant fiber traditionally used to make the sails, lines, and caulk for schooners, and performed the best in ocean trials held in 2015 by students at the East Boothbay Nature Center and at the Hurricane Island Center for Science and Leadership. Two species of fungi were tested, one a trimitic white rot of wood, and the other, a bright orange pigmented wood decomposer. The longevity of untreated mushroom buoys using the former species, ranged from three to five months in the ocean trials. The addition of one of dozens of sealants tested prolonged the life of the buoys to five to eight months.

However, no coating was ideal in addressing the criteria for lobster buoys. Criteria shared with me by the members of the Maine Lobster Advisory Council and the Commissioner of the Department of Marine Resources included resistance to ultraviolet light, physical impacts against boat hulls and decks, water penetration, submersion pressure to thirty meters, and surface biofilm fouling. The most promising waterproof coating to address these performance criteria was a PVC flexible plastic, the same one used to coat swimming pool toys and life vests. PVC, another plastic, would not do.

Unable to meet these criteria with existing partially biobased sealants, I paused the research until a time when an environmentally compatible product was developed.

Six years later, the promise of MycoBuoys derived from mycelium grown on fine hemp particles finally re-emerged. Sometimes it just takes patience to let the fungal mycelial network find the right people to play along. Recently, I got news from a chemist in Japan with whom I had collaborated in the past. He called to let me know he had developed a 100 percent saltwater- and impact-resistant, biobased, and nontoxic sealant with no Volatile Organic Compounds (VOC). The new coating kept the buoys waterproofed for two years in a saltwater tank in his lab.

Within a day or two, I also had a call from Severine von Tscharner Fleming in Pembroke, Maine inviting me to lead workshops at Smithereen Farm on how to grow MycoBuoys.[10] The application was to be for oyster and kelp aquaculture this time, and I let her know that mushroom buoys were in development. It seemed that the universe was answering my wish to bring sustainable, compostable MycoBuoys forward as one solution to replacing plastics in ocean farming.

I am thrilled to be working with Greenhorns, Seaweed Commons, Smithereen Farm, Long Cove Sea Farm, and North Haven Oysters to bring the research and development of mushroom foam flotation devices to Maine aquafarmers. Two workshops held at Smithereen Farm in 2022 focused on assembling new elongated pontoon forms to fill with Ecovative's mycelium on hemp to grow nine MycoBuoys

for deployment this season on floating Brooks Mill oyster cages.. The experiment this season will test an uncoated control MycoBuoy against three replicates of buoys coated with the new nontoxic sealant, a mixture of pine tar, linseed oil, and a commercial waterproof sealant. In addition, Abigail Barrows at Long Cove Sea Farm will test a rectangular MycoBuoy against an exact cork replica.

Our hopes for the near future are modifications of buoy shapes and sizes to meet the needs of stationary kelp and oyster seasonal aquaculture. With success, Smithereen Farm plans to equip a facility to use locally sourced hemp fiber as a feedstock for a locally sourced fungal strain as a regional business to serve the community on water as flotation devices, and subsequently, after the MycoBuoys are retired at the end of the season, on land as garden mulch and compost. Such an endeavor brings algae and fungi back to the sea where they evolved together. The seasonal exchange of algae, fungi, and plants between farming the sea and farming the land makes for a seaworthy yarn to explain a new circular economy. ○

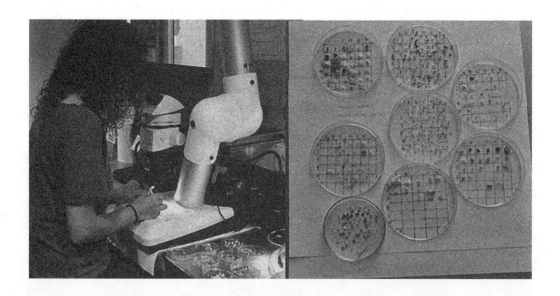

Islands in the Sea, NOAA/OER *The large aluminum frame of the neuston net is one meter high and three meters long. Here, the net is being deployed off the starboard side of the R/V Seward Johnson*

Notes

1. Darlene Duggan, "Lobster Buoys," *What I Learned at the Beach* (blog), October 21, 2020.

2. "Lobster Buoys Mark the Spot," Lobster-Anywhere, *lobsteranywhere.com/new-england-style/maine-lobster-buoys/*

3. "Better Living Through Chemistry," Wikipedia, last modified February 9, 2022, 23:28.

4. Paul Stamets, "The Petroleum Problem," *Fungi Perfecti* (blog), June 3, 2010.

5. "DART® (Deep-Ocean Assessment and Reporting of Tsunamis)," NOAA Center for Tsunami Research, National Oceanic and Atmospheric Administration, *nctr.pmel.noaa.gov/Dart/*

6. A.P.W. Barrows, C.A. Neumann, M.L. Bergera, and S.D. Shaw, "Grab vs. Neuston Tow Net: a Microplastic sampling performance Comparison and Possible Advances in the Field," *Analytical Methods* 9, (March 2017): 1446–1453.

7. André R.A. Lima, Guilherme V.B. Ferreira, Abigail P.W. Barrows, Katie S. Christiansen, Gregg Treinish, and Michelle C. Toshack, "Global Patterns for the Spatial Distribution of Floating Microfibers: Arctic Ocean as a Potential Accumulation Zone," *Journal of Hazardous Materials* 403 (February 2021): 1016.

8. Xiao-Xia Zhou, Shuai He, Yan Gao, Hai-Yan Chi, Du-Jia Wang, Ze-Chen Li, and Bing Yan, "Quantitative Analysis of Polystyrene and Poly (methyl methacrylate) Nanoplastics in Tissues of Aquatic Animals," *Environmental Science Technologies* 55, no. 5 (March 2021): 3032–3040.

9. Yaru Han, Fei Lian, Zhenggao Xiao, Shiguo Gua, Xuesong Caoa, Zhenyu Wang and Baoshan Xing, "Potential Toxicity of Nanoplastics to Fish and Aquatic Invertebrates: Current Understanding, Mechanistic Interpretation, and Meta-Analysis," *Journal of Hazardous Materials* 427 (April 2022): 1060.

10. Sue Van Hook, "Mycofoam Buoys" (lecture delivered at Greenhorns' Reversing Hall, Pembroke, ME, 2021), *youtube.com/watch?v=SsQdaVpivqg*.

Causeway

Bridges over troubled water,

radical climate adaptations,

waiting for water

Lori Rotenberk

ON HENRY'S FARM

Experimenting with Radical Adaptation to the Climate Crisis

Lori Rotenberk *Henry on the farm*

In late September 2021, Henry Brockman kneeled in the field harvesting dried beans, their vines entwined over parched, dusty soil marked by fissures resembling lightning strands. He tugged the deep brown pods in a bushel basket alongside him beneath the early autumn sun in the rolling glacial hills of Congerville, Illinois. It had been two months since it had rained.

By late October, heavy rains arrived, but the month would historically come and go without a frost. Thin ice crystals would finally blanket Brockman's operation, called Henry's Farm, in the dark, crisp morning hours of November 2. As Thanksgiving neared, the fields slipped into a slumber.

Fifty-seven year old Brockman is a small compact man who says he always carries the aroma of the last thing he harvested, will not rest, however. After twenty-eight years as an organic vegetable farmer, he says climate change has forced him to start over and spurred him to a state of constant experimentation as he works to keep his farm afloat and make it as resilient as possible for the coming generation.

Central Illinois is seeing weather and temperature extremes, as is the rest of the country. The climate there is changing more rapidly than it has in the past, explains Don Wuebbles, professor of Atmospheric Science at the University of Illinois at Urbana-Champaign.

However, the region's most significant changes are occurring at night. While the average temperature in the state has increased one to two degrees over the last century, the increase in overnight temperatures has exceeded three degrees in some parts of the state.

This part of the Corn Belt has seen a 10 percent increase in precipitation over the last century and the number of two-inch rain days in Illinois has soared 40 percent in that time. As a result, the state has seen an increase in soil moisture. But, due to elevated rates of evaporation, the soil also tends to dry out faster, and longer dry spells have become more common.[1]

All of these changes will likely make crops more susceptible to weeds, pests, and diseases, which will likely lower yields.

"We have models that say by mid-century, there will be a 10 to 50 percent decrease in yields in Central Illinois. We really have to think ahead," Wuebbles warns.

Six years ago, Brockman took a year off farming and started doing just that. He penned an emotional eighteen-page letter to his children, Asa, Aozora, and Kazami. It was written as if from the future, to warn them about climate change. Dotting it are words

Terra Brockman *Soil*

Terra Brockman *Soil holding water*

such as "mourn," "woe," and "destruction" as well as "hope."

In the scenario he imagined, it's 2050 and the farm is much smaller and mostly feeds the family. Brockman is in his eighties and corporate farms produce much less food. Their synthetic, petroleum-dependent fertilizers, herbicides, insecticides, and fungicides are no longer widely available or affordable. Combines lay rusting in nearly barren fields. The climate in Congerville feels like Dallas.

"The world is changed utterly," the letter reads, "by human avarice and ignorance, and continues to change at a rate unprecedented in human history."

Since returning to the farm, Brockman has been dedicating himself to preparing for the future. An award-winning documentary, *Seasons of Change On Henry's Farm*, documented some of the challenges he faced before and during his leave. The changes since then, he says, are coming much more rapidly and more severe.

A Small but Mighty Farm

Brockman grew up with five siblings on a one-acre farm surrounded by nearly fifty acres of woods. All the Brockmans write and have published books about farming and farm life. Three of them farm nearby and all four operations promote their work together online under the name Brockman Family Farms. Congerville is considered a mecca for organic farming due to its rich soil and rolling topography of ridges and plains formed by glaciers. The area is dotted with dozens of small-scale organic farms, some that belong to Apostolic church members who live off the land.

Outside Congerville, however, central Illinois is generally farmed by immense soy and corn operations that stretch, "a hundred miles in all directions," says Brockman. When he crosses paths with the large-scale conventional farmers who have known him since high school, he says they often call out, "Hey Henry! How's the garden growing?"

At twenty acres, he jokes his farm "isn't even large enough for them to drive a tractor onto."

Although small, Henry's Farm is lush, producing one to four tons a week of more than seven hundred vegetable varieties. Brockman farms with intention and instructs his interns to be present and respect everything they harvest, while being mindful that it will be someone's food.

"Plants, like any living thing, play their role in the cycle of life and death," Brockman says.

The farm feeds 345 families with its own community-supported agriculture (CSA) program and others in the Chicago region. It sells most of what it grows to dedicated customers at a farmers market in Evanston, Illinois, whose customers Brockman has fed for the length of his career. It's nearly a seven-hour drive, roundtrip.

For the last twenty-five years, Brockman has been taking meticulous notes in ten-year black journals, and it's there that he began unwittingly recording the mercurial weather and climate patterns he's seeing more of now. Every day he logs the weather, his plantings and harvests, and other observations such as the departure of some insects—the yellow-striped armyworm and the new arrival of others—the brown-striped armyworm. Around a decade ago, his own data revealed that things

Terra Brockman *Henry's ten-year journal*

were starting to go terribly awry. Brockman dubbed it "global weirding."

His farming season now lasts a month longer than it did when he started, extended two weeks earlier in the spring and concluding two weeks later in the fall. Springtime begins warmer but tends to be punctuated by unexpected bitter frosts that often wipe out newly planted seedlings. And rain?

"Now floods can come in the spring, in July, September, and even in December," he says. "Due to a flood in July last year, I had almost nothing at the market for a couple of weeks. When plants are underwater for more than 24 hours, their roots can't breathe and they suffocate."

Traditionally, Brockman planted heat-loving sweet potatoes in July. He now plants them in early spring and they sprout in May. His spring lettuce season has been cut short but now, due to warmer fall weather, he can plant lettuce again in the early fall and harvest it in November.

"September is what August used to be, and when freezes come they come harder," Brockman says. "I'd usually be harvesting peppers the second week of July. Now, I'm still harvesting peppers in October."

While he harvested peppers this fall, the owners of Cook Farm, a mere twenty miles south, flooded after three nearby tributaries rose and converged, leaving the Cooks to kayak across their fields. That week, between October 24 and 30, four and a half inches of rain fell onto Henry's Farm. "Somehow," Brockman said, "we had no flooding."

Preparing for the Climate Crisis

In an effort to adapt and anticipate the changes ahead, Brockman has made some radical changes to his farm. He cut production in half—he's now farming only ten of his twenty acres. A creek divides his two beloved bottomland fields, and, in an effort to save the rich, two-thousand-year-old soil on that land (and prepare for potential flooding), he moved his annual row crops to rented higher ground two years ago.

In their place, he planted an experimental perennial forest using a permaculture approach. There, he's growing currants and a variety of berries—gooseberries, honey berries, elderberries—as well as hazelnuts, pecans, pawpaws, and persimmons, in hopes that their more permanent roots will prevent the soil from washing away during heavy rains and keep more carbon in the soil. Among other experimental crops are paddy and dryland rice varieties.

Behind the forest, Brockman is growing a field of sorghum-sudangrass, which will be used as straw mulch to keep the ground moist during dry spells. It will also add carbon to the soil as he cuts out tillage and grows more with cover crops in his rotations. A bit higher in the second bottomland field is a mix of perennial grasses and legumes, as well as potatoes and garlic.

Brockman is also trying to prepare for drought. Henry's Farm rests on a deep aquifer, and the water beneath was trapped by the glaciers under hundreds of feet of clay. In other words, water there is a nonrenewable resource.

When he began farming, Brockman says he only needed to irrigate his crops in late July and early August. The rest of the year, rain was relatively predictable. In recent years, he has had to use drip irrigation beginning in May, and the need often extends into the fall. This involves running yards of licorice-like hoses dotted with pinprick holes. They are laid across seeded ground to help the plants germinate.

It's not clear whether one of Brockman's children will take over the farm, but the word "retire" isn't in his current vocabulary, said his sister Terra.

"Henry isn't focused on passing the farm on to someone else, just on farming in the best ways possible for as long as possible," she said.

His daughter, Aozora, is the only one currently working alongside her father. The author of two poetry chapbooks, she often describes life on the farm in her work. In the poem "Roots", she writes:

> In the hottest part of the day
> we sit in the shade of the shed
> in a circle of square bins
> peeling the Russian Reds

> each dirt-covered strip
> revealing streaks the color of sky
> in the last rumble before rain.[2]

This fall, it was so hot and dry on the farm that Brockman had to irrigate his young crops at night, using the water from his well. He would move his irrigation lines before dark and then get up in the middle of the night to move them again five hours later. Brockman, his fingertips stained from the tomato plants he had trellised earlier in the day, manually laid out forty lines—each two hundred feet long. As he did so, he would jog back and forth in the dark to ensure it got done quickly—so the seeds would germinate. When he finished, he made his way home and back to bed, where he says he's only rarely able to fall back asleep these days. As the year draws to a close, 2050 looms on the horizon. ○

A version of this profile with the same title was initially published on Civil Eats, and was printed with permission.

Notes

1. D. Wuebbles, J. Angel, K. Petersen, and A.M. Lemke, (Eds.), "2021: An Assessment of the Impacts of Climate Change in Illinois," The Nature Conservancy, Illinois, 4–6.
2. Aozora Brockman, *Memory of a Girl* (Durham: Backbone Press, 2016), 21–22.

Rose Linke

EBBINGS

```
where        lands
waters          meet
the               rivers
swollen             but
beds             of
grasses              tall
but           drowned
dreams               in
each            fern
itself             unfurls
bulbs           turn
on            themselves
all                 day
for               watch
falling         rocks
clock              the
strikes               out
to                 nothing
see               here
even           not
a                  stone
is            erosion
water             but
is                  erosion
also             air
```

This poem was prompted by the concept of a boustrophedon: a text meant to be read from right to left and then from left to right in alternating lines. Like a tide dancing with a moon. The word boustrophedon comes from Greek, meaning: the way an ox turns. It was once a popular form for inscribing text on stone. The poem is meant to be read in this same manner, from right to left and then from left to right in alternating lines. The ox turns, the oxbow meanders, the memory of water is held in the stone, and vice versa. ○

Amory Abbott *Flood 3*

Emily Vogler

ACCOMMODATING UNCERTAINTY

Nomadism and the Coastal Commons

Amory Abbott   *Flood 1*

On September 21, 1938, one of the most destructive, powerful, and deadly hurricanes to ever strike southern New England made landfall.[1] In addition to causing widespread devastation to cities like Providence and Newport, the storm destroyed the village of East Beach in Westport, Massachusetts. The East Beach community was constructed on a barrier beach that extended between the calm tidal salt marshes of the Westport River and the waves of Buzzards Bay. The small village included a store, a church, a post office, 120 houses, a dancing hall, and a bowling alley.[2] Every building in the village was destroyed during the hurricane.

Following the 1938 hurricane, East Beach was subsequently hit by hurricanes in 1944 and 1954, at which point the town of the town of Westport made the decision not to allow homeowners to reconstruct permanent structures and houses within the area. It was before the 1988 Stafford Act, which commits the federal government to pay 75 percent of the cost of rebuilding roads, bridges, and other damaged infrastructure after a natural disaster.[3] It was also before the National Flood Insurance Program was created in 1968, offering premiums below market rates for homeowners in flood prone areas.[4] Homeowners impacted by the hurricane may not have had the resources or insurance to rebuild and the federal juggernaut that subsidizes rebuilding in flood prone areas was not yet in place to bail them out.

The Rules of Resettling

Since then, Westport has maintained the deed restrictions that prevent permanent structures from being constructed on the barrier beach. In 2022, on a bitter cold January day, I walked down the road that was once the spine of the East Beach village. The cold wind blew off the bay and picked up anything that wasn't frozen to the vacant gravel lots that line each side of the road. It was a peculiar scene: decks without buildings, lonely auxiliary sheds, fences interspersed demarcating property lines and providing privacy to vacant spaces, and metal tent structures with no shade awnings casting long shadows on the gravel.

During the winter, East Beach Road is quiet, except for the sound of the waves breaking and the occasional sandpiper or seagull that passes overhead. Missing are the eighty or so trailers that reside here in the summer, most waiting at a farm a mile away where they get stored in the off-season. Come May, the road slowly gets reinhabited as the trailers return to their lots to match up to their decks, electric boxes, and

fences. Residents start walking the street greeting neighbors and friends as the community comes back to life.

What Can We Learn from East Beach?

Climate models and scientific reports increasingly show that tens of millions of people in the United States and around the world will be flooded by sea level rise, many places annually.[5] How to prepare and plan for storms, and rebuild after, are increasingly important challenges. East Beach presents a case study to explore questions about climate uncertainty, accommodation, retreat, and resilience. While not to be misinterpreted as a solution, it still provides a valuable lesson for how to accommodate and adapt to the migrating landscapes of the Anthropocene through nomadic resilience and the coastal commons.

Nomadic Resilience:
Inhabiting Migrating Landscapes

Climate change threatens many of our permanent settlements: inundated coastal cities, desiccated Sun Belt communities, forest towns engulfed in wildfires. Facing increasing natural disasters, communities often have to make the hard decision to resist, accomodate or retreat.

Within coastal environments, resistance includes shoreline armoring through the construction of dikes, seawalls, bulkheads, and elevating land surfaces.[6] Accommodation can mean developing coping strategies that enable continued human inhabitation in spite of increased hazards.[7] Retreat includes moving people and structures away from the coastal edge, allowing wetlands and beaches to migrate as seas move inland, and preventing new construction in vulnerable areas.[8] Depending on the specific coastal dynamics, in some places it is often only a matter of time until retreat is the only remaining option.

Underlying these decisions is an important ethical question: which communities will be protected, and where will people be given the option, or forced, to retreat? Existing data

shows that it is often class, race, and political power that answer this question.[9] These are very challenging emotional and financial decisions for individuals and communities to make. There are some communities where people want their properties to be bought out but the resources don't exist to do so. There are other places where people are resistant to letting go of their homes and communities. The concept of home and attachment to place is deep in many people's individual psychology, sense of identity and belonging.[10] While there are the financial losses associated with retreat, there are also the emotional losses of leaving a cherished home, landscape, and community behind.

Various forms of nomadism, pastoralism, and transhumance have been strategies used by Indigenous communities throughout history to accommodate and adapt to extreme climates and access to resources.[11] Whether migrating in the desert to find water, across mountainous landscapes to find good pasture for animals, or between winter and summer territories in search of food sources, communities around the world continue to live nomadically. However, colonization, capitalism, and the commodification and privatization of land has led to the enclosure of the commons globally, making this form of nomadism increasingly difficult.[12]

The East Beach community represents a form of contemporary nomadism that can increase resilience within the context of climate uncertainty. This model can allow for individuals to remain connected to a place and their community, but move out of harm's way during risky seasons. While this model is predicated on the assumption that people

↖ ↑ **Unknown photographer** *Before and After the Hurricane of September 193*
(image courtesy of Westport Historical Society)

WESTPORT

Circles indicate cottages that were on Horseneck's East Beach—at top rim of photo—prior to hurricane which wreaked havoc with Westport's extensive Summer colonies.

can afford a second home, there may be ways to adjust the model to adapt to other contexts. For example, could this form of seasonal habitation allow for a transition for communities that are resistant to retreat? Could the property be used by previous community members at no cost but rented to others to help subsidize the buyout program or to help fund the maintenance of the commons? Architecturally, it presents a unique opportunity to explore modular housing that is semi-nomadic but still rooted in a specific local vernacular architecture that reinforces the regional sense of place while providing new ways to be in community.

Claiming the Coastal Commons as a Space of Resilience

Rivers, coastlines, and oceans are among the last remaining resources still held in common by the people. The commons can be defined as something shared and protected by a community, to be kept open and protected for future generations. Within coastal areas, the Public Trust Doctrine guarantees right of access to tidal waters and shores.[13] Along the coastline in Massachusetts, where East Beach is located, the land below the low water line is considered a commons and the public has access to the wet beach for hunting, fishing, and navigation.[14]

As coastal residents face erosion, they often try to resist it by hardening the coastal edges. This creates a defined edge between the land and water, turning what was a gradient—from wet to dry and land and water—into a hard line. This depletes the rich intertidal zone where the majority of ocean life exists. It also leads to the loss of public access to the coastal commons by removing the intertidal beach zone.

↑ **Unknown photographer** *Site of same cottages (above) after the storm.*
(image courtesy of Westport Historical Society)

Harry Bloomingdale *Upper left: Bloomingdale, Trafford and other familiar property before the storm. Lower left: Site of same cottages (above) after the storm*

Horseneck Beach Right: East Beach Horseneck. Lower Right: Digging out the road on East Beach (image courtesy of Westport Historical Society)

Exactly where the boundary exists that demarcates our watery commons from the privatized land should be re-examined, as that line continues to fluctuate with the uncertainty of climate change. As sea levels rise, tide lines migrate landwards, properties erode beyond repair, and the increasing number of weather disasters make rebuilding in flood prone areas economically and practically impossible. This shifting boundary presents an opportunity to reimagine and reclaim the coastal commons as a political, social, and ecological strategy for building coastal resilience. Beyond just asserting existing laws—infringed upon every time a property owner hardens their coastline— we can go one step further and expand the commons along our coastlines to create a zone of resilience, a space where we make room for the floods. A space that allows for the species that have been squeezed between the rising waters and the fixed permanence of human settlement to be given space. A space that reasserts public access to our water commons and fights back against the enclosure of the coastal commons. And possibly, a space that can be designed to allow for seasonal use by the communities that once lived there—a space of memory, a space to return to, a space to gather.

The Future Is Unknown, and That Is Okay

The East Beach case study suggests the possibility to accommodate a shifting coast line while still allowing for a certain level of human settlement. It is a model of adaptation and accommodation that accepts and embraces community attachment to place and the desire to live in coastal environments. It suggests a temporal way of thinking about settlement that responds to the reality and uncertainty of the Anthropocene. It may help lead to an easing into the need for retreat and the associated emotional and psychological challenges of these decisions. It allows us to think seasonally about how landscapes can be used and inhabited differently in the summer and winter, and how they can be designed to support human use during part of the year and habitat and recreation during the off-season. This model also provides clues for how we may inhabit other landscapes subject to the uncertainty of climate change—the seasonal retreat from landscapes in California that are fire prone in the summer months, for example, or the creation of the resilience hydrocommons along rivers that are prone to flood.

If there is anything we have learned during the past few years of COVID-19, increased climate disasters, and political upheaval, it is how to inhabit a space of great uncertainty. It can be debilitating to make decisions within so many unknowns but it can also spark immense creativity, as we are forced to reimagine life as we know it. In the coming decades, many more of the things we take for granted will be questioned. It will require flexibility and the ability to be opportunistic in the face of change to assert what is best for humans and the more-than-human species that depend on this planet. Within this space of uncertainty, rather than reinforcing our failing systems, there is an opportunity to think creatively to move past the binaries such as resist | retreat, public | private, nature | human, land | water. This model suggests an opportunity to shift our mindsets from one of rigidity to one of flexibility. It could allow us to find peace with climate change, uncertainty, and the shifting landscapes to release our desire to control—with the understanding and acceptance that nothing is permanent, but we can still call it home. ○

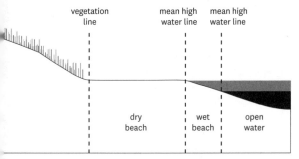

Emily Vogler *Zones of beach determine public access*

Notes

1. National Weather Service Boston, "The Great New England Hurricane of 1938," *weather.gov/box/1938hurricane*.
2. Westport Historical Society, "East Beach Before the 1938 Hurricane," 2013, *wpthistory.org/east-beach-before-the-1938-hurricane*.
3. Theodoric Meyer, "Four Ways the Government Subsidizes Risky Coastal Rebuilding," *ProPublica*, June 19, 2013.
4. Meyer.
5. Mathew E. Hauer, Jason M. Evans, and Deepak R. Mishra, "Millions Projected to Be at Risk from Sea Level Rise in the Continental United States," *Nature Climate Change 6* (July 2016): 691–95.
6. James G. Titus, *Rolling Easements* (Washington, D.C.: Climate Ready Estuaries Program, US EPA, 2011).
7. Titus.
8. Titus.
9. Karen M'Closkey and Keith VanDerSys, "For Whom Do We Account in Climate Adaptation," in Carolyn Kousky, Billy Fleming, and Alan M. Berger, eds., *A Blueprint for Coastal Adaptation: Uniting Design, Economics, and Policy* (Washington, D.C.: Island Press, 2021).
10. Vanessa A. Masterson, Richard C. Stedman, Johan Enqvist, Maria Tengö, Matteo Giusti, Darin Wahl, and Uno Svedin, "The Contribution of Sense of Place to Social-Ecological Systems Research: a Review and Research Agenda," *Ecology & Society 22*, no. 1 (2017).
11. Roger Blench, "You Can't Go Home Again: Pastoralism in the New Millennium," (London, UK: Overseas Development Institute, 2001).
12. Blench.
13. The public trust doctrine can be traced to Roman Law and is the basis for public trust laws and doctrines in the United States. It is the legal principle that asserts that the public has the right to certain natural and cultural resources and that it is the government's responsibility to maintain these resources for the public's use. See *coastalreview.org/2016/09/public-trust-doctrine-owns-beach/14*.
14. In some surrounding states such as Rhode Island, Connecticut, New Hampshire, and New York, this public ownership (not just access) extends to the wet beach below the high water line. In some states such as Washington, Louisiana, and Hawaii coastal commons include wet and dry beaches. See Titus.

Casey O'Neill

SWEETNESS OF RAIN

I awake to the sweetness of a soft and gentle rain falling on the roof. Rain has taken on much significance in these days of drought, something that I once took for granted, now treasured and rejoiced upon arrival. The rain softens me, transforming a place grown hard and brittle into one of abundance and happiness.

There was a time when I didn't covet cloudy days, sadness, and missing the sun. On sunny days, I'm still bouncier, but I've come to respect wet days in a new way. The long dry spell this winter was unlike anything I have ever encountered. I am changed by it.

I am no stranger to warm weather during the winter here in the Mediterranean climate of Mendocino County in Northern California, but, when most of the winter is such, a sense of urgency presses upon me, a feeling that I am behind in my work. When March creeps along, I'm reminded that any journey, and especially that of farming, must be done a step at a time.

And though we still need more rain, the small series of storms we've had recently has slowed things down some.

Shutting off the irrigation has been helpful, and is in a time of year when I would be thinking of nothing much of it at all yet. Prepping beds and pulling weeds are much easier when the soil is soft, unlike the hardened ground I encountered in February.

This is a year of changing methodologies for us on the farm. We are starting cannabis seeds a month later than usual, and will be getting clones later as well. This takes the pressure off of spring preparation, so that instead of several marathon days, I'm plugging away at a more measured pace, working to prep three to five beds each week as crops edge towards readiness in their trays in the hoop house.

A slower start gives the cover crop more time to grow, and also takes pressure off of us in a year without extra labor to help with a big spring push. After months of being on the waiting list, we just got shallions and green garlic. I had hoped to plant them earlier in the winter for an earlier spring harvest, but they will still provide flavor and variety for the market sometime in the early summer.

Shallions are fun because they are the sprouted shallots leftover from last year that would otherwise go to waste. I plant the bulbs with the green tops sticking upwards, and in a couple months they will be harvestable as bunching shallots, similar to green onions. Green garlic is much the same, making use of sprouted heads to become future flavor instead of going to the compost pile.

We've been harvesting the big winter brassica from the hoop houses, which has been crucial to fill our market table and community-supported agriculture (CSA) program, and is also now freeing up space for the next plantings. More beets, bok choy, salad mix, turnips and radishes are going in, marching out through the shifting kaleidoscope of crop patterns that make up our winter rotations.

The big brassicas were a late gamble to fill up planting space after the fall harvest. We didn't have anything else available to plant, and were too burnt out to be able to do anything more than pop in some starts and see if they would do their thing. Overall, the large brassicas are better suited to life outside the hoop house, while quick rotation crops are more effective at maximizing the high-value space in such sheltered locations. There have been some problems with rotting in the cauliflowers and Romanescos, which have denser heads, and the cabbages have been super slow to head up.

As a result of our experiments, we're making plans for low tunnels to shelter the brassicas for cold snaps by dropping the plastic over them, but avoiding the lack of air movement and overheating that happened during the hot, dry periods of this winter. Shifting them to the big terraces on the lower part of the farm will allow us to make better use of the winter hoop houses in what is becoming a strong offering for winter markets and CSA.

It's always tricky to balance winter farming with burnout, and we're also trying to have the conversations about self-care, and about not pushing so hard during each different season so that we maintain some reserves of energy. We learn and grow as time passes, and, on this rainy morning, I take a moment to treasure the ride. ○

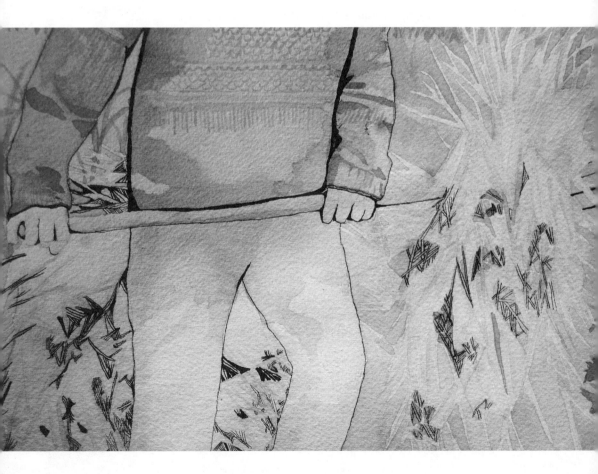

Jacob W. Forquer

FARMING IN THE FIRST PART OF THE DAY

She gets dressed in the dark, lingers and wades around the room finding the things she needs for her day: a trowel dusty with dried mud, her boots similar, a handkerchief flowing from her basket. She leaves before me, having nurture to tend to, things covered in frost, peppery and green. As the day warms in the rising sun, frost retires to dew and she manufactures lemon balm by rubbing wet grass in her hands, where dirt has pushed itself between her finger and the skin of her silver and stone ring. The turquoise stud returns to the earth in breaths like a swimmer. In the working morning, I see her room blush with a rosy bloom; feel stoned smelling orange blossoms. I walk home after making the bed with a very fine quilt on top. When she returns, her basket is full with three heads of lettuce, basil and carrots curled around each other like legs on a picnic blanket.

Vincent Medina and Louis Trevino

CAFE OHLONE

A Love Song to Ohlone Culture

Cafe Ohlone, the only Ohlone restaurant in the world, is a love song to our beautiful culture that is Indigenous to the San Francisco and Monterey Bay region along the Central California coast. Every element of Cafe Ohlone speaks to specific elements of our identity and was created as a tangible way to provide representation for our living community in our urban homeland, while simultaneously teaching the public—through delicious Ohlone foods— of our continuous presence in the East Bay. Located at the University of California,

Berkeley, it's an environment where culture can be promoted in a dignified manner, and a space where increased consciousness is fostered with our specific Indigenous identity. Please let us take a moment to walk you through.

••

Area I: A Portal of Light and Shadows
The first area upon entry pays respect to our original Cafe Ohlone space at University Press Books, where the cafe in the back was hidden from view to diners entering the bookstore. This added a layer of mystery and wonder to the experience as, like with a portal, guests were transported into a world reimagined. After entering, diners pass through a scrim-covered, open air series of shadow boxes framed by redwood. This area is illuminated by the light of the sun during the day and lit softly at night, displaying dramatic silhouettes of native plants dancing in the wind, altering shape and movement with the changing light. The plants showcase our traditional East Bay landscape and traditional Ohlone foods, as well as subtly introduce Cafe Ohlone's native plant gardens. Diners get their first view of the burlapped native gardens after walking along a path from this first area. The gardens through-

Cafe Ohlone *Quail eggs*

out Cafe Ohlone are full of aromatic and greatly respected plants that are made into teas and added to the plates of our meals.

Our voices are a part of the landscape in Cafe Ohlone's singing trees. The sounds diners hear upon entering Cafe Ohlone, amplifying as they reach the second and third areas, come from native trees singing in the Chochenyo language. In our old-time beliefs, still held close by Ohlone people today, there was a time before humans existed when plants could speak. This dreamy paradigm inspired Cafe Ohlone's singing trees: six native trees, each with the voice of a living Chochenyo speaker. An elder's voice begins to sing by herself—a familiar oldie, a ballad heard in our grandparents' homes. This beautiful, wise voice is joined, one voice after another, with voices of different generations, until all are singing in unison, in harmony. All the trees appear to be singing at once, saturating Cafe Ohlone with song in our enduring language. Then, each voice drops off, one after another, until only a child is singing in the language, representing the next generation of Ohlone culture and those still to come. When the trees complete their song, the voices shift to jokes and conversation in language. Birdsongs of geese, flickers, blue jays, and Red-winged Blackbirds serenade diners when the Chochenyo voices take a break.

Area II: A Dry Creek Runs Along Redwoods

Diners arrive at the first of two dining areas. On one side of the courtyard is a dry creek made of Napa Basalt rock, interspersed with riparian plants. This area features a series of dining tables, each handmade with locally and sustainably sourced redwood.

Most of the dining seats are made out of redwood logs, designed with comfort and sustainability in mind. Along the dry creek, two dining pods use boulders as seats. The entire space is speckled in abalone and traditional baskets are displayed with great respect. The native plant gardens continue throughout this area, and the air is perfumed with Indian flowers and medicinal plants.

Area III: A Shellmound Rises in a Fragrant Garden of Abundance

The lavish gardens culminate in the third area, centered around a long, communal redwood table for our Ohlone elders. The wall behind the family table is trellised with native blackberry vines; a redwood and tule ramada—a traditional shade shelter—covers the communal table. Abalone hanging from the ramada's roof twinkle in the evening candlelight.

The largest of the singing trees is here, as well as a storybook oak tree—also with a Chochenyo voice—in recognition of the importance of oaks and acorns in our Ohlone culture. The southern side of the area features a new shellmound: a raised hill made of crushed oyster, mussel, and abalone shells and earth that pays respect to the traditional Ohlone shellmounds that ring San Francisco Bay—

sacred sites for our people, built by the hands of our ancestors.

The dining areas feature an abundance of native plants such as artemisia, poppies, mugwort, ceanothus, hummingbird sage, and Indian strawberries—much as the landscape was prior to colonization: full, yet managed.

•••

Cafe Ohlone is a one-of-a-kind culinary and educational experience. Every meal of luxurious Ohlone cuisine educates the public about our rich, enduring culture. Our menu has expanded, with sustainable sources of traditional foods added by season: fine black oak acorn soup; bay nut truffles; Ohlone salads; San Francisco Bay Dungeness crab and king salmon; seared venison backstrap; Tomales Bay mussels and clams; Olympia oysters, the native oysters to this part of the world; chanterelle mushrooms; Indian strawberries; Indian teas made of elderberry, hummingbird sage, yerba buena, and rosehip. Each meal is accompanied by a description of the specific context of our beautiful Ohlone culture to better educate the public about our living culture.

Cafe Ohlone, is a tangible cultural space for our living Ohlone community; a space where language classes are held, safe gatherings and meals for our elders occur, and where our community can see representation of our cultural identity outside of our homes. It is also a permanent center in the effort to build relationships with departments throughout campus to better relations between Ohlone people and the University of California, Berkeley.

For so long, Ohlone people have lacked these physical spaces within our homeland. Cafe Ohlone is centered on creating a safe space for our community to be represented and made visible. We acknowledge, and are fully aware of the harm the University of California, Berkeley—specifically, the Hearst Museum of Anthropology—has historically caused to our Ohlone community. We are taught from our elders that Phoebe Hearst, the founder of the Hearst Museum, had a positive and amicable relationship with our family in the late 1800s and the early 1900s. However, we also know that after her passing, both her son and the University had a role in the loss of our peoples' federal recognition after Alfred Kroeber, director of the Hearst Museum from 1925 to 1946, erroneously wrote that Ohlone people were "extinct for all practical purposes" in 1925, despite the fact that he was aware our people continued to persevere.[1] Kroeber's words lended to the forced removal of our family from the Sunol Rancheria in 1927, an act in which Phoebe Hearst's son, William Randolph Hearst, had an active role that is recounted by our elders. Around that same time,

the University removed our ancestors from their cemeteries, and they continue to be kept at the Hearst Museum against our peoples' wishes. Our role on campus is an effort to uplift the truth, to foster healing, and to develop new, better relations that will propel us into a more respectful, just future.

We acknowledge our undeniably strong ancestors and wise elders, who make it possible for our Ohlone culture to continue on. Their constant work allows for the blossoming of culture we are witnessing today. In the Chochenyo language, the first language of the inner East Bay, we say holše mak-nuunu—our culture is beautiful; tuyye mak-muwékma—our people are strong; ḥinšušte mak-miččiya—our elders are wise; makkin nommo mak-warép hemmen tuuxi—we've been here in our home all along; 'at makkin rootesin nommo mak-huššištak 'ayye—we'll be here tomorrow, too. ○

Note

1. Sabrina Imbler. "New DNA Analysis Supports an Unrecognized Tribe's Ancient Roots in California." *The New York Times*, published April 12, 2022, nytimes.com/2022/04/12/science/ muwekma-ohlone-tribe-california-dna. html.

Veronica C. Nehasil

HUMAN ECOLOGY

Talismans

In the late spring of my second year of college, I was given the gift of seeds. In the dim light of a warm barn in central Maine, I listened to Mohawk and Wabanaki elders and white folks share stories about corn. I was introduced to the underland again, subterranean rivers of history, of more buried pain.

When I left there, I had a small bag of golden corn kernels in my pocket. I hadn't wanted to take them at first; I don't have any land. But I brought them home to Michigan that summer, built a raised bed with my mother, and we dug. The corn grew straight, tall, and green. The broad-bladed leaves fluttered in the humidity. The cobs developed: hundreds of sleeping lives tight in their perfect rows.

Months later I sat on the wicker couch in the dampness of the greenhouse beside Suzanne, my botany professor. Rain drummed on the glass roof in the twilight of the early autumn darkness. It was two days before a final presentation and I still hadn't grasped the existence of transposons; gene expression in the aleurone layer of individual kernels of maize that morph during development until they're sealed in suspended animation, like a bee in amber. The existence of transposons creates unique pigmentation in each fruit—in fact, many different pigments can be expressed in the same fruit because of these genes. She was patient with me, using her hands to show me how the genes jumped, the layers of lipid, protein, and carbohydrate cells which created the seed, and the layer in which these genes could only exist. Layers of seed like layers of earth.

The underland yields what is buried. What goes down must come up—eventually, transformed. Genes in microscopic layers of seed bloom themselves into a borealis of color and the garden has grown. It's produced ears of golden, black, smoky blue, and rose maize; and dozens of swollen bean pods, inflorescences of burgundy amaranth as tall as the kitchen windows, winding squash vines which trail onto the lawn, blooms of marigold from Mesoamerica, zinnia, sunflower. Their seeds, living capsules of sheltered genetic memory, talismans. ○

Hannah Althea

NATURE'S NATURAL BUILDERS AND COB CURIOSITY

In the fire-ridden West, the smell of campfires trigger sense memories of orange summer skies filled with smoke and remnants of habitat loss, human and animal alike: Evacuating and praying your home isn't gobbled up. Hungry fire fueled by a terrifying mix of gales and drought. These raging fires are increasing in frequency, with months-long burn seasons in California anticipated as early as February and as late as October.

To be resilient in our shifting climate and landscapes, we have to get creative. For that, we can turn to the beaver.

Beavers are excellent natural builders. Using sticks, mud, and grass, they build effective dams in the bodies of water where they live. These structures have an integral role within their ecosystem—offering water filtration and preventing erosion—and their mud-stick architecture seems to harbor vegetation from wildfires.[1,2] Scientists have observed regions of Colorado where landscapes were ravaged by fire, the areas stewarded by beavers were often still green and untouched. These landscapes—ponds with marshy surrounds, connected by streams cleverly created by the beavers' architecture—somehow allow them to be left unscathed by fire. It's not clear why.

Where I live is nestled half a mile from the Willamette River in Portland, Oregon, at what was once colloquially deemed the Planet Repair Institute (PRI). It serves as an urban permaculture learning container dedicated to sustainable living against the odds of cultural values that espouse capitalist resource consumption above all else. In its energetic heights, it has had various flourishing garden beds. Greywater systems! Aquaponics! Urban perma-culture design and natural building workshops! Abundant herbs! And houses many passionate activists by being able to purposefully offer affordable rent! I was lucky to stumble upon this home three years ago when searching for a room on Craigslist.

The land itself is home to a cob structure (named the Sanctuary) offered up for passing guests—and other structures enhanced by the beauty and insulative yet breathable nature of cob. The Sanctuary is ostensibly the first known cob structure to be erected in Portland, about twenty years ago. Cob—a mix of clay, sand, and straw—is strong, fireproof, and resistant to impacts of seismic movement. It is generally affordable, accessible, and easy to learn. Natural builders will tell you that it's hard work and can be slow, but lends to quieter working sites, and a lower barrier to entry

Suzanne Husky *Smokey the Beaver*

organic materials for pigmentation, they mimic the yellows, browns, reds, blues, and pinks that can be found in the elderberry plant or the roadside buttercup.

When the sun is out, the window-filled Sanctuary glistens. It has timber-framed windows and shelves to prominently display ceramic treasures collected over the years. A built-in bench extends from the walls on either side, with enough space for a cot.

Embedded within one of the cob walls is a "truth window" purposefully exposing the straw bale insulation in a framed circle, to celebrate the alternative insulation instead of toxic fiberglass insulation in most buildings today. To the left of the Sanctuary, a small pond with goldfish and aquatic plants is lined with a stone wall. Every inch of wall, floor, and dust reminds that, just like you, it was once of this earth, but will outlive us all.

Natural building is slow by modern building standards, a slowness that is not rewarded in our ramped-up capitalist system. Maybe this slowness, too, is an offering for another path. Beavers follow rhythms that tell them to eat, to build, to sleep, to gnaw. The seasons unfold slowly around them. They are informed by the sound of rushing water alerting them to a leak to tend to, or the turning of leaves reminding them to store food in their lodge for the upcoming winter.[4] The synthetic rhythms built under capitalism call for quickness, but at the cost of leaving each other behind: underhoused, underfed, and ecosystems imbalanced and plundered.

Shifting community priorities at PRI means that the permaculture systems on the land I live on still function, just quietly. The gardens await their starts after a winter of mulching. "Weeds" in the shape of comfrey and lemon balm threaten to overtake the garden beds. An

to learning the craft. And it is always, always, worth the beautiful result. The exact origins of the building technique are not certain, but its results are widespread worldwide. Earthen homes are found in nearly every continent you can find human dwellings: Europe, Asia, Africa, and North America; and many have lasted centuries.[3]

When you enter earthen spaces, the legacy of cob can be felt in your nervous system. The fire-resistant and long standing structures are effortlessly beautiful. Because they are usually literally built from the ground up and the mud-straw-sand mixture can vary, they can be innumerably different in size, texture, pigment, and shape. Round corners are shaped by hand, and because of the use of

Hannah Althea *Nature's Natural Builders: Beavers home at Errol Heights Wetlands in Portland, Oregon*

out-of-place banana leaf tree needing pruning looms green and bright among a spring of rainstorms. Nettles will be harvested in March as comfrey and underground creatures are growing wildly and wriggly in an empty garden bed. When I look out the window of my home, the Sanctuary heartbeat of PRI still stands strong and beautiful, a daily reminder in the form of an invitation for other ways to be: when in community, in living spaces, in deciding how to build within our environments. I am still learning.

Modern beavers build meticulously and cleverly, often out of sight from the human eye. Evidence can be found though in the gnawed branches and logs strewn across the marshy dirt, felled trees, the dams raising the water table to new heights, and the mound in the middle of the pond that is the beavers' earthen home. I don't claim to offer solutions here, just curiosity: Can natural building offer a resilient and adaptable future rooted in justice? I think we should ask the beavers. ○

Hannah Althea *Nature's Natural Builders: Cob "Sanctuary" at Planet Repair Institute in Portland, Oregon*

Notes

1. Alex Hager. "To Improve Wildfire Resistance, Researchers Look to Beavers." N P R, October 16, 2021. *npr. org/2021/10/16/1046779486/to-improve-wildfire-resistance-researchers-look-to-beavers.*
2. Emily Fairfax and Andrew Whittle. "Smokey the Beaver: Beaver Dammed Riparian Corridors Stay Green during Wildfire throughout the Western United States." Ecological Applications 30, no. 8 (2020). *doi.org/10.1002/eap.2225.*
3. "Section 1: Introduction to Cob." *Oregon Cob & Beyond (blog). University of Oregon Blogs. December 2015. blogs. uoregon.edu/oregoncob/history-of-cob.*
4. W. Zurowski. "Building Activity of Beavers." *Acta Theriologica* 4, vol. 37, *infona.pl/resource/bwmeta1.element. agro-article-e3dd2b1b-7198-4b0e-ba62-fd0335984ec8.*

Mike Iocona and Scott Kessel

REFUGE YURT

This Refuge Yurt has a twenty-foot diameter at the eaves and is intended as a simple low-cost shelter that can be built with friends in a matter of weeks. You don't need carpentry experience to build it—but a willingness to try, make mistakes, and to try again will help a lot. It's a beautiful thing to play a part in making one's own yurt, and to be sheltered from wind, rain, and snow.

This yurt is air sealed but not insulated, which means it'll warm up nicely but won't hold heat as long as an insulated yurt during the coldest days of winter. Insulating the roof adds a little complexity, but can certainly be done. Woodstoves are a simple, elegant heat source and wood is inexpensive or free in most regions. But don't skimp when installing your pipe—an improperly installed chimney has reduced many a yurt to a pile of ashes. Windows are luxurious and can be added at any point, but the skylight featured in this design alone provides ample illumination.

The walls and roof of this simple tapered-wall yurt are made from fifteen hundred board feet of kiln or air-dried boards. The floor in this design has a sixteen-foot diameter and it consists of just under eight hundred board feet in the girders and joists. It is supported by only four foundation points to save digging and allow for easy leveling of a concrete-free foundation. This ensures that your egg will always stay in the center of your cast iron pan. The flooring and roofing styles are up to preference. Two-by-six tongue and groove boards make a stout floor, but it can be challenging

to source them locally. A one-inch pine shiplap or tongue and groove floor can also work, as long as the joists are spaced to match. A cedar shingle roof is unparalleled in beauty, but has a shorter life in a wooded or damp location. Metal lasts a long time and is great for collecting rainwater, but falls short on aesthetics. Tempered glass table tops are easily acquired, inexpensive, and serve perfectly as a skylight.

This tapered-wall yurt can be built with one or two people, but it's usually more fun to have a crew. Spread the word—you'll be surprised whose ears will perk up at the chance to help raise a yurt. Or, consider hosting a workshop to help inspire more people to build yurts. Yurt raising workshops are held with the intention of spreading the infectious fascination and great appreciation for the wisdom of the nomadic peoples of the Asian Steppes who developed this shelter technology we call a yurt. There are many ways to build a yurt and the design is constantly evolving. What insights and delights present themselves to you in the process of building your yurt? ○

Mike Iocona and Scott Kessel *Refuge Yurt*

Nobody to Serve Coffee

Surplus and bounty,

shifts,

changing energetic priorities

Barbara Rose

REWILDING TASTE BUDS AND NEIGHBORHOODS

Sonoran Desert Food Forests Nourish Community

In 1985, my family, new to the Southwest, was offered a caretaking position on twenty acres in the foothills of the Tucson Mountains of southern Arizona. Rural before our time there, its inter-mountain wildlife corridor was facing intense development pressure as Tucson grew. As caretakers, we kept cattle, horses, sport shooters, and artifact hunters off the prickly, saguaro-studded land owned by an eighty-five-year-old woman who had lived here many years before. Her stories showed us how much this place meant to her, even after she moved to California to be with her daughter. In an unlikely turn of events, we soon became owners as well as caretakers.

As the original people of this place have always known, the Sonoran Desert is a food forest where Indigenous peoples thrived for thousands of years before colonization, resilient and replete with delicious, nutritious, and drought-resistant legume trees; fruiting understory and herb plants; insects; and animals. Relatives. Their descendants continue to educate, repair, celebrate, and advocate for acknowledgement and reparation. Early on, we began studying and practicing permaculture, first taught by Bill Mollison on the lands of the Tohono O'Odham Nation. This practice has connected many diverse people and

organizations in the community: writers, artists, builders, herbalists, gardeners, teachers, dreamers, farmers, water harvesters, wild-crafters, and activists.

What grew from many years of place-based study and practice is Bean Tree Farm, a wild foods farm and learning center. The decades-long journey to become of this place has enabled us to cultivate strong and enduring relationships, as well as learn to build and share knowledge about water-wise desert homes, neighborhoods, and gardens. The farm's name seemed to easily characterize the forests of ironwood, palo verde and mesquite. These trees produce protein-rich beans and pods; provide shade; and nourish cactus species (pads, buds, and fruit which hydrate and regulate blood sugar), desert berries, herbs, and understory plants, which all provide superbly healthy and healing foods—a Sonoran Desert cornucopia! That these foods are so delicious and nutritious is even more impressive when you realize that these plants thrive on about twelve inches of rain per year, and have been through severe drought cycles over time.

Our farm has been grateful for opportunities to partner and collaborate with Desert Harvesters and the Community Food Bank of Southern Arizona to build economic, social,

Barbara Rose *Farmhouse and lower section of the farm*

and food justice through hands-on education for desert dwellers about rewilding urban and suburban neighborhoods with Sonoran Desert food forests. Colonizers have controlled, oppressed, appropriated, and fragmented local native cultures and landscapes for centuries, and we believe our community and farm can have a role to play in helping to create positive change. In December 2021, Desert Harvesters was selected for a Community Food Bank Thriving Communities' Grant that will fund a video series, a new desert foods cookbook edition, and a partnership with a traditional Tohono O'Odham harvest and education group who work with Indigenous youth. Alongside city and university sustainability departments, Bean Tree Farm is a grant partner, linking a wide array of grassroots, academic, and municipal efforts to address food, water, climate, and justice matters. ○

Barbara Rose *Sprouting ironwood and palo verde seeds with dry mesquite pods*

Robert Dash *Fava Cover Crop*

Ingrid Ellison

MARCH

Nothing is green but the lime in my kitchen
Coveting bright acidity under its prehistoric skin

I could thatch the lawn but the mud would win
So I am on my knees
Pulling back leaf litter in the beds
Looking for spring
In the form of the tiniest bit of color
Sword tips, shark fins, chartreuse fighting their way
Through the deep muck

Back in the kitchen I look for more green to assemble on the counter
A bottle of pills for the dog, a child's tea cup, a produce bag
A bit of lichen from the windowsill, loopy green elastic

I open the refrigerator and pull out everything green
Celery, cucumbers, kale, a granny smith apple, the lime

I heave the juicer onto the counter
The thing that has too many parts to clean
But I take it out anyway
And begin chopping a mosaic of textures
Fibrous, leather, glossy, crisp

I turn on the whir and plunge the pieces down to the sharpest blades in the house
And watch the steady foamy stream pour into a mason jar

I drink all of it
Swallowing in gulps
Sending a green shiver into my chest

Fallen Fruit (David Burns and Austin Young)

A MONUMENT TO SHARING

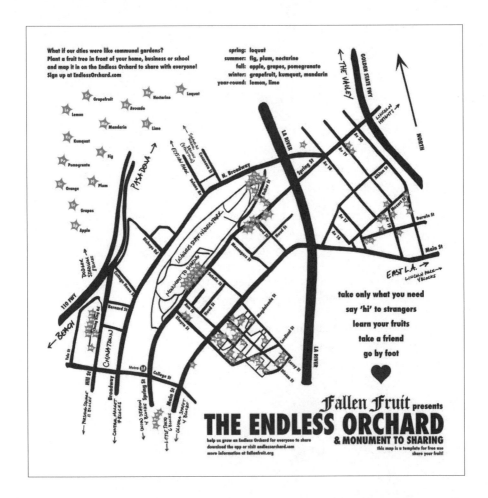

A Monument to Sharing consists of thirty-two orange trees in planters located near the Ann Street Entrance of Los Angeles State Historic Park. Each planter has a quote from a neighbor who lives nearby. They become a poem of collective voices that wrap around a public orange grove and lead to fruit trees planted within the community. ○

BE SUPER PRESENT AND FEEL THE AIR ON YOUR SKIN AND
Do you remember what dirt, grass and flowers tastes like?   Smell
Walk around the neighborhood, don't drive a car. Turn off
YOUR FEET ON THE PAVEMENT.   LISTEN TO THE SOUNDS
the fragrance of the high deserts on the Santa Ana winds. Say "¡por
your technology and pay attention to everything happening
AROUND YOU AND SMELL THE AIR.   LISTEN TO BIRDS SING
favor!" when a woman yells "¡Tamales!" Run out to the ice cream
around you. Look at everyone you walk by. Look people
-ING TO EACH OTHER.   NOTICE ALL OF THE CRACKS IN THE
truck when you hear 'Its a Small World' playing. Grow tomatoes
in the eye, smile and say hello. Ask a question, like "are
SIDEWALKS. SLOW CHILDREN AT PLAY. PICK HONEYSUCKLE
and basil in your flower beds. When you pass a plum tree, an
you having a good day?" Make a friend. Offer a stranger a
FLOWERS AND SUCK THE SWEETNESS FROM THEM. NOTICE THE
apple blossom tree, a cherry tree, a lilac bush, a field of wild Ti-
piece of fruit. Pick up any trash you find. Leave only kind-
COLORS OF HOUSES YOU PASS BY. COUNT THE NUMBER OF
ger Lilies, or a freshwater spring that appears out of nowhere, give
ness. Look out for something you haven't noticed before.
PEOPLE WHO ARE RIDING BIKES. HOW MANY FOR SALE SIGNS
them all special names. If the snow is plowed and makes a cor-
Trees are residents of a neighborhood longer than most
ARE IN THE NEIGHBORHOOD? IS THERE A COMMUNITY GAR-
ridor on either side, just follow the opening. If the fog is so thick
people who live there. Imagine the neighborhood be-
DEN? HOW MANY BARKING DOGS CAN YOU COUNT? DO YOU
you can't see the end of the block - just keep walking! Look at all
fore there were people or streets or houses. If someone
SEE ANY CATS? CAN YOU SPOT A BUTTERFLY? WHAT DOES
the beautiful doors and stoops while imagining who and what has
has a fruit tree that they are not picking, knock on their
YOUR NEIGHBORHOOD TASTE LIKE? DO YOU SEE ANY BEES?
happened there over all these years. Spend more time looking
door and ask if it's ok to pick it. Notice the different types
LISTEN TO THE WIND. LISTEN TO YOUR FOOTSTEPS. MAKE
up than looking down. How many for "stop signs" are in the neigh-
of fences in front of the buildings...each one is a mark-
STREET ART. PUT A BASKET OF FRUIT FROM YOUR BACKYARD
borhood? Bring an apple pie or casserole to the new neighbors.
er of how far away you are from home... Hang out on
IN FRONT WITH A SIGN THAT SAYS "FRUIT IS FOR SHARING"
When the streets lights come on, you had better be running home.
front porches and sit and gab with neighbors for hours.

Fallen Fruit *A Monument to Sharing (Collage)*

THE FRUIT TREES IN THIS PARK BELONG TO THE PUBLIC

THEY ARE SHARED AND CARED FOR BY THE COMMUNITY

WHEN THE FRUIT IS RIPE GENTLY PICK WHAT YOU NE

AND LEAVE SOME FOR OTHERS.

Fallen Fruit   *A Monument to Sharing (Victorville)*

Fallen Fruit   *A Monument to Sharing (Orchard)*

Emily Gaetano

MY CLAY-COLORED MOTHERLESSNESS

Green fronds waving against blue sky, rust-colored clay mountains burning in the distant haze, crimson blood on white paper. My rude awakening that something was terribly wrong was set against an idyllic backdrop.

I became a Peace Corps agriculture volunteer in Cameroon, West Africa, after college graduation in 2017. My self-designed degree focused on sustainable farming and food studies, and working with the organization felt like the logical next step toward a career in agriculture. In Cameroon, I lived in the rural village of Banefo-Mifi, where I immersed myself in the community's local traditions and farming practices. Working alongside farmers and local leaders to improve nutrition in the village, I was taught to plant beans while following the contours of the land, and I introduced new techniques to produce natural fertilizer by making compost. I was humbled to be accepted into this community, and to be able to exchange agricultural knowledge and skills.

Nine months into my volunteer service, I used the bathroom and saw blood. Tests revealed that an amoebic infection in my intestinal tract had caused chronic bleeding and twisted gut. Clean drinking water is often unavailable in the area, so amoebas aren't uncommon. My body was not used to the bacteria, though, and it took three rounds of antibiotics to finally eradicate them. By then it was too late. I was medically separated from the Peace Corps at the beginning of 2019, when a colonoscopy revealed that the amoebas had triggered a full-blown flare of ulcerative colitis, a disease that causes large, painful ulcers to form in the colon.

I was devastated by my diagnosis. It was hard to accept that I was truly sick, and to abandon my community and our farming plans for the next year. The rug was pulled out from under my feet and I found myself back in my parents' house, jobless and torn from the life I had expected to keep living.

I was determined to carry on with my life and career aspirations, though, and that spring I started working with a local landscaper. I loved being outside with my hands in the dirt, but by the summer my still-failing health forced me to hang up my tools. Every few minutes I was doubled over in pain from intense cramps, and my intestinal bleeding—worsening, even months later—made me so lethargic and anemic that lifting a trowel to uproot a weed was a strain. The next few months were a blur of hospitalizations, blood transfusions, and rounds of antibiotics—even supplemental, intravenous nutrition and a myriad of food-exclusion diets. I had been a pescetarian for over a decade, but I was so desperate to increase my iron levels that I ate ethically raised meat from a local farm, whose humane and transparent practices assuaged any guilt I'd previously felt about partaking. These trips to the farm, through meandering countryside, were some of the few moments of tranquility and normalcy during my period of illness. I was able to share that time with my mom. But despite these medical interventions and nutritional sacrifices, everything I ate made my colon feel as if it was being turned inside out. Surgery was my only option for survival. In October 2019, I received a total colectomy, a surgery in which my colon, or large

intestine, was completely removed and replaced with an ileostomy.

I was physically and emotionally trauma-tized from my year of illness, but my body quickly healed from surgery. Soon, my energy came back, just in time for a new career opportunity with the National Park Service in Washington, D.C., working at a greenhouse that grew plants for the White House. Yet another stepping stone in my farming and flowers career path, this job provided the perfect mix of hands-in-dirt work and federal benefits. I looked forward to a long and stable career scurrying around the greenhouses.

But before I started, just months after my own surgery, my mom was diagnosed with amyotrophic lateral sclerosis, or bulbar onset

Futurefarmers *Open Akker* (photography by Caroline Vincart)

ALS, a fatal neurological disease which is marked by nerve death and muscle weakness. My weeks took on a steady routine of watering, fertilizing, and pruning plants in the nation's capital; my weekends were spent driving to and from Connecticut, spending time with my parents by helping them move into their new house, which was being remodeled to accommodate the electric wheelchair that would soon become my mom's only mode of transportation.

Her medical care needs quickly increased, and I joined my father and brother as a caretaker in January 2021. I permanently left my job to do so. The rapid progression of my mom's illness stripped away her ability to enjoy many of life's simple pleasures, and I saw all of her post-career plans get quickly thrown away as her health declined: first losing her voice, ability to eat, then the strength to walk or move. I spent most of my time doing health-related tasks for my mom, such as preparing medication, helping her shower and use the bathroom, or calling the doctor when a new symptom began, but I also planted a perennial flower garden in my parents' new front yard. I was brought to tears when my mom said the garden made the house finally feel like home, and I was happy that this garden provided all of us with a beautiful and serene respite from the daily demands of being a patient or a caregiver. Despite receiving the best at-home care, my mom passed away in July 2021, about eighteen months after being diagnosed with ALS, and three months before her sixtieth birthday, which is when she planned to retire.

My mom was a proud Navy veteran, and she was working for the United States Department of Veterans Affairs at the time of her diagnosis. When her symptoms became too much to manage at the office, she took medical retirement— a bittersweet way to end a career in which she was valued for her work helping others. A relaxing retirement had been just ahead for my mom, and she had eagerly anticipated countless hours in the garden, quilting, sewing, and visiting her friends around the globe. ALS took this away from her.

My unanticipated separation from the Peace Corps shook my sense of security, and my mom's ALS diagnosis showed me that nothing in life is certain. Her premature death at fifty-nine years old forced me to rethink my perceptions of meaningful work, my expectations and assumptions about life, and how I want to spend my future days knowing that they aren't truly guaranteed.

My love of the natural world is likely a result of the countless hours spent in the garden with my mom. Farming has been a part of my life since childhood. I believed that land work would constitute my career path, and that perhaps I'd even become a farmer. Now, I feel myself drawn away from agriculture and toward an artistic career. Although gardening has been a source of comfort and recovery the past year, wonderful memories of crafting with my mom have brought much joy and

relief as well. My mom kindled my artistic creativity and craftsmanship, too, and I see my new burgeoning career in textile arts as a chance to promote awareness of our respective diseases. I believe my means of bringing good to the world will be as an artist, and that I can give more this way than as a farmer. I hope my art can ease the suffering and alienation related to illness and grief, and that it will also be an integral part of my personal healing. I am firm in my new conviction to pursue art.

Still, I have a vision of agriculture's future: regardless of a person's job title, we are all more connected to our food and to the people who grow it. We can prove to farmers of future generations that there is a demand for their work by supporting those who practice regenerative farming and sustainable land stewardship. Farmers are the reason we have food on our tables, and I still want farming to be an integral part of my life—and I believe many people feel the same.

Cultivation and creation are inseparable to me. I'm collecting materials in my garden; silver scissors flash and another hibiscus flower is separated from its stem. These fuchsia petals will transform into fabric dye for my next art project, a giant pink colon suspended from the ceiling... ○

Poki Piottin

THE SWAN SONG OF A YOUNG ELDER

I walked into the field ready for harvest and the Red-winged Blackbirds were there, chomping on my Chihuahua blue corn. For the past two weeks, I had yelled at the happy fellows for pissing me off by eating the fruits of my labor, and was ready to again. But instead of a scream, my heart opened to a sense of gratitude for the ability to feed wildlife.

I worked a two-acre field pretty much alone last summer, growing blue corn, bolita beans and winter squash, waiting for volunteers to come farm in community like our ancestors had done. I spent several days with a young Hopi man who shared songs, stories, and wisdom from his elders, like "tend your corn every day as if they are your children." I seeded by hand with a couple friends following me to close the furrows, laid four-and-a-half miles of drip lines, and was able to get germination before the irrigation ditch went dry.

Hail, monsoon rains, and winds knocked over the corn several times. Hilling the fallen stalks, weeding, and the heat wore me out. I took long naps in between hoeing sessions. All other farm work was put on hold and I went into "management mode," as one of my elders recommended.

A handful of friends came to weed, but I harvested, husked the corn and winnowed the beans pretty much alone for ten weeks, yielding fifteen hundred pounds of corn, five hundred pounds of squash and four hundred of beans. After six years of primarily large-scale urban farming and four years on this farm, last summer was a brutal initiation. I felt a great sense of accomplishment, but I was also exhausted. Not having young people learning by my side brought me great sadness.

I started this nonprofit farm along the Pecos River, ninety miles from Sante Fe, New Mexico, after running a popular two-acre urban farm in the city. I built a campground. I became commissioner of the irrigation association managing our thirteen-mile-long ditch, wrote grants to teach kids the culture of the *acequia*, the communal care of the irrigation ditch. Several essays I wrote on the topic of repopulating farmland were published in well-read regional papers. I spent the winter alone, making gallons of soups with the crops I grew, and resumed construction on a farm building I established when I moved here.

Locals were invited to come share space, as were various permaculture and garden groups online. I repeatedly invited my previous urban farm audience to come participate in the simple chores of a farmstead and explore the potential of this farm as a community building

Poki Piottin *Farm*

center. A resident made ice cream and cacao brownies for visitors. But to my great dismay, hardly anybody came. Instead, people cheered me up on Facebook. I oddly felt like the lone star of a Survivor-type reality show.

In March 2022, I turned sixty-four years old, and had to admit that my strategy—"if you build it, they will come"—had failed. Continuously waiting for people to participate in a community endeavor didn't make sense. I know that my experience is not unique, nor is it related to my age. It is a phenomenon of our times that we need to address because the wisdom of the elders is being wasted when opportunities are offered and not taken.

So it is with sadness and a sense of relief that I am putting the property up for sale. I am pleased with what I have accomplished and learned while stewarding this beautiful piece of land. The trees, berry bushes, and cover crops I planted—and the food I grew—fed a lot of wildlife. I gave myself wholeheartedly to both the project and the local community. I have no regrets. The nonprofit was created as a community land trust, with a mission of exploring the creation of a permanent agrarian settlement, one of the principle tenets of permaculture. Most people who idealize a back-to-the-land lifestyle don't seem to understand that affordable farmland isn't

Esha Chiocchio *Poki at the* acequia

Poki Piottin *Sunflower*

located at the edge of a city, and that the practice of tending the commons and growing a surplus of food, another tenet of permaculture, is necessary. All of this takes great effort, resources, courage, discipline, commitment, and ingenuity. If it was easy, there would be plenty of flourishing agrarian communities in the countryside.

I lost ten pounds during the summer, but, since I ate almost all the ice cream and brownies we had made for visitors, I am back to my normal weight and ready for my next adventures as a young elder. I have been able to take in all the beauty that I've created here, and begin a healing process that was long overdue. My wish is that the next owner of this farm benefits from all the work I did and finds a way to preserve agricultural land for future generations. I believe we must create new communal settlements, where farmland is cared for, food is grown, children are raised, and people are cared for through sickness and death. ○

Sharon Stewart

EL AGUA ES LA VIDA

Water is Life

In a region of ghosted presences, situated at a double oxbow bend in the Pecos River, the remote and tiny El Cerrito has endured for centuries. This survey presents a village life portrait animated by interdependence on water from the community irrigation channel, the *acequia*.

No one remembers hearing of its origins, lending to speculation that the waterway was created by Indigenous peoples who farmed along the Pecos.[1] Others believe its existence can be ascribed, as numerous others of New Mexico's seven-hundred-plus acequias, to the efforts of Franciscan priests, who, when colonizing the region, established two vital elements of village life—water and faith. Very likely a confluence of efforts set this hand-dug gravity flow irrigation system that traces the village.[2]

Acequia also refers to an association of users that honors water as a community resource rather than a commodity. This is anchored in the Islamic Law of Thirst, which says that all beings have unfettered access to water and it never be hoarded or sold. The arid farming irrigation methods that originated in the Indus Valley, and later the Arabian Peninsula, were then developed across North Africa and brought to Spain with the Moors and their seven-century occupation.[3] Sharing in abundance and scarcity is a vital tenet of acequias, which are also a self-governance system. As parciantes, members of the acequia hold water rights and elect a *mayordomo* as caretaker to oversee maintenance throughout the year—especially during the annual spring *limpia*, or cleaning.

In El Cerrito, the limpia is the one social gathering outside the rare wedding and more common funeral for which extended family, friends, and curious students of traditional village life return. *Parciantes* have shared for generations the responsibility of maintaining a waterway that feeds their families, orchards, gardens, fields, and livestock. While recharging watersheds, acequias also provide a rich riparian zone for wildlife, shade trees, and native plants, many of which are used in traditional medicines.

Seeing the universal in the personal, this exploration of El Cerrito's survival provides insight into co-operative perseverance; a model redress to the dissonance created by our time's reigning individualistic ways. With the availability of water a defining issue of this century, monetary pressures are strong on rural residents to sell their water rights to transfer for urban development needs. Doing so would

Sharon Stewart   El Agua es la Vida: *Water is Life: Abran and Vidal*

Sharon Stewart   El Agua es la Vida: *Water is Life: Walking the Ditch*

Sharon Stewart   El Agua es la Vida: *Water is Life: Mayordomo* →

effectively sever ties to water and land that are a deeply cherished cultural component of this region's agrarian communities. In resisting, villagers ensure the survival of the Southwest's oldest extant water system while reinforcing the universal truism:

EL AGUA ES LA VIDA   WATER IS LIFE ○

Sharon Stewart
El Agua es la Vida: *Water is Life: Macario's Little Field*

Notes

1. Charles P. Loomis, "El Cerrito, New Mexico: A Changing Village," *New Mexico Historical Review* 33, no. 1 (1958), 53.
2. Ira G. Clark, *Water in New Mexico: A History of Its Management and Use* (Albuquerque: University of New Mexico Press, 1987), 9.
3. Jacquetta Hawkes, *The First Great Civilizations: Life in Mesopotamia, the Indus Valley, and Egypt,* (New York: Knopf, 1973).

Zev Stephen York

TAKING

An Ode to Sugaring

metal bucket bound,
i race arms outstretched into Sugar Maple.
listen! just this morning,
before i bent my knees in reverence,
lowering my eyes to a tree-blood-coaxing metal spile
i heard an unaccompanied tap tap tap of downy woodpecker:
today's rhythm missing a syncopated drip drip drip of sap.
looking up, i felt stillness fall over cold air.

singing my vibrations through frozen heartwood,
i remember their movement:
last week's sap rising until buckets brimmed with blessings
of springtime.
today, I'm aware of their
pause.
not in breath, but nonetheless in flow:
sugary food
climbing
to engorge buds
rests
for a moment
in a chilly root cellar.

did paw paw seeds not evolve delicious fruit
to hitch rides in gentle, sugar-drunk human hands?
sometimes, even summer's greening sun lies cold in Sugar Maple veins,
so just one minute longer under this canopy of twigs i'll rest.

returning home to pancakes,
days of vigorous movement glaze my plate:
i taste sunshine and hear rhythms of Sugar Maple's drip,
memories of warmth clumsily drizzling over my bottom lip.
yes, i may be taking
a share of last summer's sunshine.
but as i run my eyes along Sugar Maple skin,
knowing of pause in their own innergowing,
i can't help but feel as if this sort of taking
binds all of us further
in the sweetness of syrupy rapture.

Rimona Eskayo *Springstorm*

Sensory Configurations

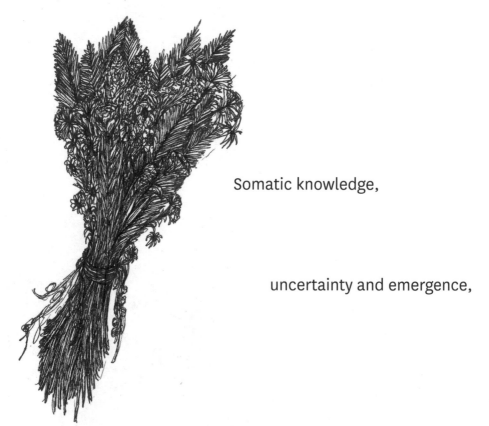

Somatic knowledge,

uncertainty and emergence,

planting adaptations

Renée Rhodes and Alli Maloney

LETTING THE LAND LEAD

An Interview with Nance Klehm

Nance Klehm *Nance in prairie grasses*

Nance Klehm is a radical ecologist whose work explores the intersections of soil, community, health, and ecosystem wholeness. She works to reimagine waste streams, soil contamination and fertility, and agroecological practices. Nance leads community soil science trainings and workshops, is a landscape designer and horticultural consultant, and has written books such as *The Soil Keepers: Interviews with Practitioners on the Ground Beneath Our Feet*. After years of centering her work in urban places she bought a multi-acre property called Pachamanka in the Driftless Area, where she is working to build health and connectivity.

What drew you to the land you're currently stewarding at Pachamanka?
I had been looking for land for fourteen years and had just settled on a smaller piece in central Illinois that had only a few trees on ten acres, surrounded by an ocean of corn and soy. I went to an Earth Day celebration and learned that they were going to start mining fracking sand from the state park very near this land. I went home and cried in the bathtub and decided to walk away from the deal. Someone suggested I attend a land auction that was happening. I went and, in just over eight minutes, I became the owner of this land, Pachamanka. It was a sideswipe, although looking for land was my intention because I grew up rural on five hundred acres. It's all been part of my ancestral story for generations.

This land is in the southernmost part of the Driftless Area, a region that stretches from Illinois into Wisconsin and Minnesota, about forty minutes drive from the Mississippi River. Illinois is known for being flat but this area is not; there's elevation change because it wasn't glaciated and thus not scraped flat by retreating glaciers. We have deep soils made by prairie ecosystems and extensive riverways and streams as well as thin soils that exist on limestone ridges. It's a mineral-rich soil that is currently dominated by farmed corn and soy, with some wheat and oat and dairy and wool operations in places where rock is too close to the surface and farming isn't possible.

What feels important to you to be doing there?
I make a punch list every morning and evening, but when I walk outside, I let the land lead what I need to do. I have a diversified growing operation, mostly perennial crops, not so many annuals: mushrooms, perennial vegetables and medicinals, fruits and nuts. I'm growing for habitat and for the greatest biodiversity, not growing with food as a goal—food will be abundant. I keep Cayuga ducks, Bobwhite quail, and honeybees, and anytime you have animals you answer their call first. I don't consider myself a farmer and I don't use the term permaculture. I use the term agroecology because it more aptly fits what I'm doing. I build soil, insect, and bird life, all that supports the health of plant fungi and keeps systems moving around. I insert projects into larger living systems and do some restoration of the woodlands and prairie here, although it's not conventional by any means.

I've been here seven seasons and actively growing here for five. There's woodland and an old timber woodlot remaking itself. I've just started working in that area in the rows of trees that would never grow together in a natural woodland like the registered forester of the forestry service designed it. It's a very biologically non-diverse part of my landscape—in fact, most unbuilt lots in Chicago have more biodiversity than this four acre woodlot.

What plant communities are present on your land? How are you participating and exchanging with them in the work that you're doing?
There are plant and fungal communities, grassland, bottomland, wetland, woodland, prairie, and recovering agricultural fields, plus this woodlot. What I'm doing now is not about identifying but understanding how everything is working together over time and extreme seasonal change. Year to year change has happened: we're heavily impacted by flooding and face colder winters. It has fluctuated widely in the five years I've been engaged.

I have been noticing weather my whole life in relation to plants because I come from a horticultural family. Our livelihood depended on it. I'm watching how plant communities work as opposed to individual plants or species. I'm recording notes on the shift, not fighting it, or trying to hold ground on it. I'm trying to go with what is happening.

We have had two back-to-back, once-in-a-hundred-year floods. And two years ago, I started losing my pond to drought which means I lost amphibian diversity, insect diversity and bird diversity. The sound scape changed. In 2020, two acres of a pond disappeared. In 2021, also a dry year, the water table dropped and other plant communities took over those formerly saturated soils.

Pond soils are conditioned to be low oxygen. A whole different plant community succeeded in the drought and successfully went to seed. I didn't see any caterpillars last year. We had a low number of fireflies. We used to have some of the best firefly viewing anywhere because half of my property is bottomlands and they like the tall grass with wet feet. It wasn't just impacting what I was trying to keep alive— everything was different.

This year we started getting the pond back and Sandhill Cranes came back for the first time in three years as a result.

What you're observing is changing what you work on or how you tend to the place in the day-to-day. Do you have a longer-term plan that you're making adjustments to as needed?
I'm trying to blur the edges of the land that's cultivated around me with the land I have. Two, arguably three, sides of the land are cultivated farmland by my neighbors. I'm interested in blurring those edges because they can be quite abrupt. I have what I would call the "hairy" property. It's a stark difference. I'm interested in how I can work transitional zones to soften that and to create edge habitat that is the most productive, biologically speaking. I'm interested in how I can work with a new diversity of tree species—new in the sense that they don't exist very much in the current landscape, but they're indigenous to this area and have largely already been milled out—different species of oaks, hickory, butternut, and pecan.

I spend a lot of time planting trees and tall shrubs. Other areas are trying to hold the line between the invasive grass that was introduced by the federal government to feed cattle back in the 1950's, by burning and mowing. I don't spray, which is what's advocated by state, county, federal, and even restoration companies. I'm much more hands-on. I don't have any children; I'm working to give this land back. I believe in the Land Back movement and I've been opening those conversations because I know they take a long time.[1] That's where it's going, back to someone, be it the Black farmers of this area, or the Indigenous.

What are some of the changes being projected for that region as the climate shifts? Are there ways of preparing the land for those shifts?

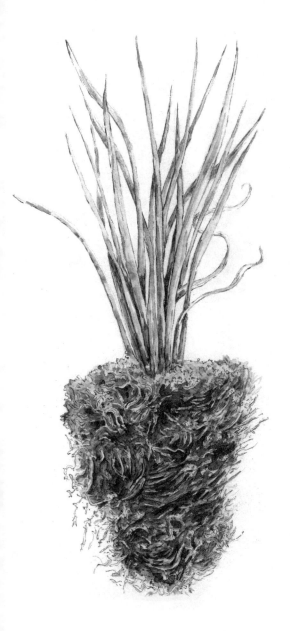

or green mulching. I work with all nitrogenous and carboniferous waste to stay on site. I'm madly collecting clean farm wastes and have tremendous compost piles. I have a diversity of processes including a lot of ferments that I'm working on and I try to marry all those processes every time I expose soil, whether I scrape it with my tractor because I bumped over something or I burn a prairie, or till, which I very rarely do. I'm always thinking about what I lay down on the surface of the soil in copious amounts. I have a layered—most people call it overgrown—highly forageable landscape, because it's not about markets or products. It's about the health of the landscape and the health of the soil.

I plant more than I need to and always plant an understory underneath all my trees. I'm not planting guilds. I'm planting prairie plants as understory in my orchard. I'm throwing down alfalfa and different clovers. It's not a perfect plan. I'm looking at broader communities: I step on those seed heads at the end of the season, I knock into them with a stick and let them fly. I'm interested in how these systems perpetuate with minimal, but continual, engagement—minimal in the sense that I'm not overthinking anything, but I'm continuing to engage a little bit more like an animal than a planning person.

What is your perspective on soil and soil health? Does it shift when you're working in a city context?
I grew up on a very diversified farm with animals. Manure piles and working with waste was something that we did because we understood waste as production and besides, where is it gonna go? I went to Washington, D.C. for school. My academic background is archaeology, which is when I really started looking at soil.

Everything's about the soil. Even if I'm planting, I'm thinking about soil, about what I'm putting in with that seedling, or with that tree, or what I'm top dressing it with. I do a lot of chopping and dropping of vegetation,

Vincent Waring *Blue Eyed Grass*

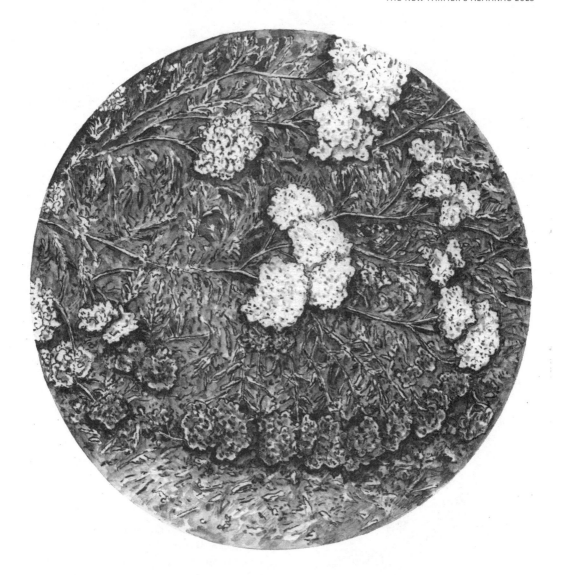

I noticed neglect: people saw themselves running across the surface and didn't see the dimensionality of it. The world goes up and the world goes down and they're just running over the surface. Most peoples' vision of their world is just whatever their body takes up. It is just real estate. As an archaeologist, I noticed that soil changes depending on what's in it. I knew that, but to see how it changed visually and chemically with different materials that were buried. I came to Chicago after graduating and doing fieldwork in coastal Peru to work at

The Field Museum of Natural History and I started getting involved in gardens because gardening was something that I did and knew. I've always been interested in all that urban crap is doing in soils and what does it mean or how it affects the people that live there—not just because of what they grow in them and then put in their mouths but because they're walking over soil's surface. In the past twenty years, I have been looking at that, hard.

No one seems to make a connection between soil health and community health. They do if you stick it in your mouth, but they don't if you just happen to be living there. Community gardens and urban growers generally don't want to know and don't have the money to test the soil. And even if they find out the imbalances and contaminations, what are they gonna do then? That's the question we are all facing. There are many challenges. I love big problems and I just have been going at it head-on ever since.

Working professionally in bioremediation, I build soils for resiliency so they best meet the challenges they face unique to their specific context. With compaction and/or repeated disturbance being the most common in any soil, I address this first and foremost. Pollutants or nutritional imbalances I dive into later. I feel like I'm a student of soil in its intersection with water. Water is something else that I spend a fair amount of time watching.

You hope for your land to be a place for community and for people to come together and engage. Are there projects or work that you're looking forward to?
There's always new ideas in the mix. I've had individuals to groups of two hundred and fifty people here for a week. It's wide-ranging and some things are retreat-based, and there's

a creative residency called "Chop Wood, Carry Water." I'm interested in the more expansive definition of creativity, outside of and beyond the art world. Some people organize groups to come here for discussions because it's a lot easier to have discussions on land than it is to have it indoors in a city where everybody runs home or is checking their phone. Here, they can do it on walks. They can do it around fires while they're camping. People come through and do long-term studies or their own about plants, animals, or growing systems. There's a school that's going to be happening, too. A lot of conversations happen out here. I believe land facilitates important conversations.

I've been developing the buildings to be of community use. There's a lot of studios and a barn kitchen, and there's a big lounge. I'm trying to make it useful to people—it's not just for growing, it's a space for gathering, too. I'm trying to build a broader flow of community with the idea that I'm gonna give this to somebody at my death, or when I'm tired.

It's hard not to think about legacy at my age. I'm fifty-six. What am I gonna do with all this, given where the world's at? I believe in food sovereignty, I'll get behind it, but that's not what I'm working on. I'm working on land health and land politics and interested in how those cycles can also interface with an agricultural movement in a healthy and powerful way. ○

Note
1. The Land Back movement is a call for the return of Native lands to Native peoples.

Linda Weintraub

TURKEY TALE

Another Other Mother

Every turkey has a tail, but few provide the content for a turkey tale. The turkey tale that follows offers an opportunity to test the innate maternal wisdom of animals. For those whose encounters with farm animals are limited to films, books, and television, even a routine report regarding animal nurseries might be news. This narrative relays four extraordinary episodes in the lives of the turkeys who lived amid pigs, sheep, chickens, ducks, and other animals on my homestead in upstate New York. It commences with egg laying, persists through hatching, extends to include the rearing of turkey chicks, and climaxes with an unanticipated exception to the caregiving and selfless attentiveness associated with motherhood.

Chapter One

The pair of turkeys I acquired as chicks matured into a proud, white-feathered tom and a sleek, full-breasted hen. In spring, the hen made what seemed to be a foolhardy choice by selecting, as her nest, an empty plastic planter where chickens had already laid six eggs. The first day there were six chicken eggs and one turkey egg; the next day, six chicken eggs and two turkey eggs—and so it progressed for six days. Then she climbed aboard, arranged herself, and settled in for the twenty-eight-day

incubation in this utterly misguided location: the planter was exposed to the rain, provided minimal shelter from the sun, offered no protection from predators, and lacked access to food and water.

I was riddled with uncertainties about whether I had an ethical duty to intervene. Did responsible husbandry entail asserting better judgment by relocating the nest? Is it presumptuous to assume I know about egg-sitting better than the turkey hen? I also fretted about removing the chicken eggs from the turkey nest. Would this be a prudent act or arrogant interference?

I opted to leave the hen, her eggs, and the chicken eggs in place. I waited and I watched. I fretted.

Chapter Two

When the late spring sun shone relentlessly, she baked in the heat. When it rained, she was drenched. Even when winds howled, she would not budge. My efforts to tempt her with water and food failed. Neighbors and friends began checking in, eager to see if she would reject the chickens as alien intruders, or take them under her wing, alongside her own offspring.

Just as the countdown to hatching was

coming to an end, the hen satisfied our curiosity. Evidence that her boundary-defying maternity had penetrated egg shells appeared one morning in the form of six dead chicken embryos. They lay on the ground at the base of the planter amid the cracked remains of their shells. This mom was no fool, and no under-achiever either. It required a remarkable feat of acrobatics to maneuver all six of the chicken eggs straight up the sides of the planter with her beak, and push them over the rim.

Six strong turkeys hatched a few days later. Displaying the attributes of Mother Goose and Mother Earth, the hen was as vigilant with her chicks as she was persistent in sitting on her nest. Meanwhile, the tom took little notice of his progeny. He strutted around the barnyard—haughty, vain, and self-absorbed. At regular intervals he spread his huge wings wide and low to the ground, raised his tail, fanned it out as wide as it could go, and strutted like a royal sovereign as the tips of his wings dragged on the ground, making a loud and threatening scraping sound. Then he would raise all his feathers, throw his head back, and jut his bill forward, causing his crop to inflate and shift from light blue and white to a vivid red. The performance culminated with deep *chump* sounds and a vibration of his tail feathers. Parental nurturing was not on his agenda.

Chapter Three

About three weeks after hatching, I discovered white, fluffy feathers strewn about the barn-yard—all that remained of the beautiful hen and five of the thriving chicks. The sole surviving chick was peeping in pitiful distress. He raced about for a while, and then fixated on his dad. The baby pursued the tom and then, remarkably, the tom followed the baby. I panicked. Once again, faith, confidence, and responsibility collided: Should I separate them because the tom might attack the poor, motherless chick? Should the dad determine its fate? Should I assert my impulse to protect the chick?

Extension Service Photographic Collection *Oregon ranch woman feeding a flock of turkeys* (image courtesy of Oregon State University Special Collections and Archives Research Center in the Commons)

What actually occurred was the one outcome I did not anticipate. The tom pursued his chick not to harm him, but to protect him. From that moment on, he relinquished his conduct as an aloof dad and adopted the demeanor of a vigilant mom. He and his chick were inseparable. At nap time, the chick crawled onto his dad's back and nestled in the soft feathers between his powerful wings. The tom would settle down and not budge until the chick awoke and scampered to the ground, resuming his pecking, peeping, and preening.

But the tom did not just fulfill maternal responsibilities for nurturing and sheltering; he underwent a total gender trait reversal. Masculine swagger vanished, along with all gobbling, displaying, and strutting. He not only assumed the caregiving duties of his departed female mate, he assumed her postures, clucking sounds, and instincts, along with her patience and vigilance.

Chapter Four

Fast forward through the summer, fall, and winter. The tom remained as attentive as a proud hen as his son grew into a fine specimen. With the return of spring, however, different instincts awoke—ones that overwhelmed parental caregiving and offspring care-receiving. The days warmed and lengthened, as did the procreative urges of father and son. With no hen on site, they sought opportunities to gobble, strut, and display their feathers beyond the borders of the barnyard. The females who inspired their seductive performances belonged to the flock of wild turkeys that frequently visited the homestead.

Again I debated: Should I confine the boys because they had become my pals, or let them fulfill their yearnings in the wild? Would reliable offerings of grains and greens, fresh

water, and clean straw persuade them to remain at home, as bachelor turkeys? At first, they circled around the fringe of the visiting flock. Later, they ventured further and began grazing alongside the uninvited guests. Then, one rainy morning, they ambled toward the woods with the wild hens and disappeared. I never saw them again.

Almost every morning now, a group of four wild turkeys visit our meadow to graze. They are not typical. Unlike wild turkeys, which are usually dark, white feathers appear on their wings and bodies.

While I miss our stately tom and his spirited son, I am comforted by the hope that these gorgeous white turkeys not only survived the deadly attack; they also survived the challenges of living in the wild. Perhaps, despite the limits of domesticated wisdom, they lived long enough to invigorate the genetic pool of wild turkeys for generations to come. Perhaps they will comprise the next chapter in this turkey tale. ○

South Carolina Department of Natural Resources
Domestic versus Wild Turkeys

Lune Ames

WHEN THE MOON IS IN THE FEET

Four years ago, my Granmom first told me about how the "sign of the Moon" oriented her upbringing on an Indiana farm. She could still hear her mom say it was time to wean the calves; she swore her mom's sweet potatoes tasted better than her dad's. Only the women in her family lived by the sign of the Moon, and the 1945 *Everson Farm Manual* offers a glimpse into how it functioned.

The zodiac is mapped on the human body:

Aries (head and face)
Taurus (the neck)
Gemini (the arms)
Cancer (the breast)
Leo (the heart)
Virgo (the bowels)
Libra (the loins)
Scorpio (the groins)
Sagittarius (the thighs)
Capricorn (the knees)
Aquarius (the legs)
Pisces (the feet)[1]

Several paragraphs document which signs are best for certain activities, one of which is weaning: *Many mothers recommend weaning babies when the sign is going downward from the knees, or in the legs and feet... This is also the time to wean calves or hogs.*[2]

Weaning is best, then, when the Moon is astrologically in Aquarius or Pisces. The Moon follows the same ecliptic, or path in the sky, as the Sun, just faster. Its position changes every few days rather than every month. To find it, Granmom described checking the *Farmer's Almanac*. Later, as I scoured all kinds of 19th- and 20th-century almanacs, I found an image called the Man of Signs with the zodiac adorning the body. That this lore was nearly forgotten in my family compelled me to dig further.

Also known as the Zodiac Man, or *Homo signorum*, the figure first appeared on a cunei-form tablet and has been a memory device to orient agriculture, medicine, and domestic life for millennia. As I researched, I kept asking what the sign of the Moon was and how it has functioned across its many histories, but I was asking into the past. I tried lunar gardening, but it felt like a performance. I surrounded myself with lunar calendars, but I couldn't figure out which Moon to follow. Maria Thun's *Biodynamic Almanac* tracks the constellations, not the astrological signs, whereas the *Old Farmer's Almanac* includes both.

The astrological sign is about one constella-tion off from where it actually is in the sky, and

the Earth's wobble is to blame. Like a spinning top, it shifts around its axis every 25,800 years—so, every 2,150 years, our celestial backdrop shifts by one constellation. I gave up imagining it all working like clockwork.[2] Some gears are missing, others are rusted. Things don't always seem to align. Some days I still check the Moon's position, and like a prayer I think, *What's being forgotten in order to be remembered?*

I decided to draw the signs onto my own body as a different clock in a different era. I collaged zodiac animals from a 16th-century German woodcut and renamed it the Moon Body. As I hung myself by the feet like the tarot's Hanged Man, I realized that the sign of the Moon is about bodies healing bodies. It's about their interconnectedness on multiple scales—the human and nonhuman, the Earth, and the solar system. "When" is a matter of relationship.

The saying is that time heals, but time is measured by conditions that are impressed upon matter. It's really that relationships heal. It's being *with*. Beyond physical healing, the Moon Body points to connection, meaning, and memory as essential medicine for the collective—ancestral, cultural, and ecological.

In forgetting how to interpret the astrological signs, their meaning changes. Though I absorb how others have historically ascribed meaning, I must make my own. Inverted, the Moon Body becomes a mirror. It's not sweet potatoes I'm planting, but myself. While everything hangs in the balance, the Moon Body doesn't arouse a feeling of certainty, but rather a faith in the unfolding. ○

Briana Waltman   *Lift Off*

Note

1. Ray D. Everson, *The Everson Farm Manual* (Huntington, IN: The Indiana Farmers Guide Publishing Company, 1945), 173.
2. "Precession of the equinoxes," *Encyclopedia Britannica*, last modified March 24, 2022, *britannica.com/science/precession-of-the-equinoxes*.

Jennifer Monson

TWO SCORES

from *A Field Guide to iLANDing: Scores for Researching Urban Ecologies*[1]

Mushroom Walk / Foraging
for any number of participants

Choose one square foot of ground.
Examine that square for an hour.

> —iLAB Residency, Strataspore, 2009
> Collaborators used mushrooms as both metaphor and material to discuss
> infrastructure, networks, and latent potential. The process cultivated "spores"
> of knowledge about local ecosystems and urban sustainability.

Future Site
for any number of participants; writing and drawing materials

Take fifteen minutes to observe and map a site.
Take five minutes to record your response to these questions:
Thinking broadly about ecological succession, what do you see succeeding here?
What is the possible future of this place given current elements?
Share your maps and notes with each other and discuss.

—iLAB Residency, *The City from a Plant's Perspective: Mapping NYC as Native Flora*, 2007
The collaborators explored how plants, landscapes, and people exist as and within
physical structures that all move, all the time. Botany, design, and choreography have
points at which their investigative methods, classification systems, and ideal outcomes
intersect, and strengthen one another. ○

↖ ↑ **Carol Padberg** *Mycelial Sleeve 1 (with wool and oyster mushrooms)*

Note
1. The scores collected in *A Field Guide to iLANDing: Scores for Researching Urban Ecology*, a book published in 2017, are both records of past interdisciplinary practice and open questions that came out of fifteen years of programs from the Interdisciplinary Laboratory for Art, Nature, and Dance, or iLAND. These two scores come from the iLAB Residency program, each residency was composed of two to eight collaborators from different disciplines researching New York City's urban ecology. Reprinted with permission from 53rd State Press in New York.

Casey Whittier, Anna Andersson, & Aubrey Streit Krug

SENSING SILPHIUM

Casey observes from her home, Kansas City, spring 2022

Many art techniques start with this directive: observe. Engaging with and describing material using my senses—touch, texture, color, shape, form, shadow—is core to my training.

I witness a plot of silphium grow from a few low leaves into a buzzing nine-foot-tall wall of dense green and yellow. From the jagged edge of one leaf to the smooth, cupped form of another leaf emerging, I sense life: resilient and disorientingly diverse. We give our daily attention to a mere sliver of what we live among, even in the most suburban landscapes. So how can we grow our awareness?

Aubrey describes the project at her workplace, central Kansas, spring 2020

I walk a familiar path into remnant prairie, wending toward a wild patch of *silphium integrifolium*. This native perennial has adapted to thrive across a wide swath of North America. Silphium's deep roots, sturdy stalks and leaves, and sunflower-bright blossoms have attracted the attention of microbes, herbivores, insects, and—most recently—humans, including me and my colleagues at The Land Institute.[1] I crouch down for a closer look.

When I stand up and lift my head from the prairie, I can see domestication research plots that include silphium. We envision a future perennial oilseed crop broadly adapted to environments made even more dynamic due to climate change. I want to grow stories of relationship with silphium. We invite a small, decentralized network of civic scientists to conserve the diversity of locally-adapted silphium ecotypes, so their seed is available for research and restoration.

Casey welcomes silphium to her home, Kansas City, spring 2020

I'm saying "yes" to civic science, but I don't know how it will shape me. I'm intrigued by the commitment to make a place for silphium, observe the plants perennially, and share information and seeds with fellow civic scientists and researchers. This is a minimum three-year commitment of time, space, and attention.

I begin with a familiar practice: making bricks from earth and reclaimed ceramic materials. I define a place for this endeavor. The physical work grounds me and foreshadows the work of planting, watering, weeding, and watching. How can I be both an amateur and expert, a civic scientist and a civic artist? How might a partnership that takes me beyond my usual sphere of influence and professional

Anna Andersson   *Casey looks up to greet some of the thirty-six silphium plants in her home garden*

Anna Andersson *A silphium rosette soon after planting in Casey's plot, lined with homemade ceramic bricks composed of recycled material from her ceramics classroom*

circles change my understanding of who I am? How can experiential community learning expand our imaginations about the future we can create? It feels invigorating and just the right amount of uncomfortable.

Anna visits Casey and silphium, Kansas City, summer 2021

On a warm July morning, I follow Casey past a menagerie of native plants lining the yard and patio into the silphium nursery nestled behind the garage. These prairie plants are a specific ecotype collected from St. Claire County in Missouri, and are only weeks away from a towering yellow bloom. Later, when the seed heads are dry and brown, Casey will

Casey Whittier *Beaded clay sculptures representing* Silphium integrifolium *flowers at different stages*

harvest the seeds of this distinct geographic population.

I review photos and data Casey has submitted so far in preparation for my visit. But within minutes of being here, I sense new data. One could only collect this data by observing a person close to their plants. As Casey introduces the plants to me and we gaze into the foliage, she rubs her finger across the leaves, noting textural differences. She senses lesions and bumps, explaining what strain of pathogens she suspects are the cause.

With her confident tone and the field-specific descriptors she uses, I wonder if I'm speaking to a plant scientist and then remind myself that she's an artist. Yet, with this practiced method of inquiry, she's at an exciting intersection of the two.

Casey reflects on the project, Kansas City, summer 2022

In my third season, silphium has redefined my yard and my relationship to it. I've developed a new way of caring for these plants. In lieu of anxiety about watering, diseases, pests or other ailments that commonly befall a garden, I see each of those phenomena and organisms as part of the complex ecosystem that silphium supports and promotes—all of it worthy of witness. I've replaced my own garden expectations for perfect blooms and lush greenery with empathy and a perennial curiosity. Civic science has given me a model for gardening that is less about control and more about witnessing.

I don't know what the results will look like. There are innumerable gifts in the practice of observing. Envisioning a different future isn't just about imagining it. It's also about the slow work of observing what is. From what is, we build plans and paths to what will be. ○

Brandon Forrest Frederick
Casey's glass wall hanging depicting Silphium integrifolium

Note

1. Founded in 1976, The Land Institute is a non-profit organization working to develop diverse, perennial grain ecosystems that produce ample food while achieving levels of ecosystem functions needed to make human life sustainable.

Margaret Wiss

UP IN THE AIR

Margaret Wiss   *Day 8*

Margaret Wiss   *Day 21*

Margaret Wiss *Day 201*

Margaret Wiss *Day 133*

Benjamin Prostine

COMPOST PILE

     Toss it on top
dusty blankets of junky hay, excrement
of various kinds: human, hen, lamb, cat,
pigeon, cow—soft bones of the broth pot,
piss-soaked shavings of wood, rotted
sheep hides, the viscera of laying hens:
lungs, kidneys, pipes, feathers, intestines—
rock hard crumbs of bread, hair, egg shells,
toilet paper, clipped toenails, menstrual
blood & a wood stove potash heap: locust,
elm, oak, burnt down scraps of poems,
bank statements, bills, newspapers, dreams—
cat fur, cabbage hearts, leftover creamed
soups gone sour, wax (ear, bee, tallow),
possum spit, cow snot, casserole's slowly
grown white fuzz mold, yellow willow leaves,
year old krauts now gray, coffee grounds,
cobwebs, hard cider moldy under the sink,
blue corn chips, mouse corpse, banana skins,
seeds of apple, flax, black cap, orange, man,
picked scabs, wine corks, quack grass, xanthan
gum, dead zinnias, violets, yarrows, cosmos,
rot of hen-of-the-woods, dryad's saddle, coral—
hail, rain, sleet, sun, ice, frost, air, snow,
elements, organs, flows, bright & pale colors
to change, decay, to become rich wholly stuff—
what's tossed on top turning good & dark
                   down below

Rimona Eskayo *Symbiosis*

Christine Heinrichs

THE OLD-FASHIONED WAY TO POULTRY SUCCESS

"Ringlet" 1st Prize and Champion Cockerel; 1st Prize Cock and 1st Prize Pen Cockerel at Madison Square Garden, New York, 1911. **The Three Greatest Living Barred Rock Males.**

The American Poultry Association's Standard of Perfection is the exhibition and small flock bible. Breeds wax and wane in popularity, but those that have stood the test of time continue to delight their keepers. Chickens are adaptable and the vitality they carry in their genes is a bridge to the future.

The Standard organizes breeds into six classes: American, Asiatic, English, Mediterranean, Continental and, the catchall, All Other Standard Breeds. Each breed has its own special history, and each class offers breeds that can serve small producers well.

Here are six representative breeds to consider: American Plymouth Rock, Asiatic Cochin, English Cornish, Mediterranean Leghorn, Continental Polish, and the Games, now in the All Other Breeds catch-all. These breeds are still leaders in small flocks, exhibitions, and the hearts of all who keep them.

Plymouth Rocks were developed in Massachusetts after the Civil War and named for one of the state's most famous landmarks. Barred Rocks are the most popular color variety. These birds are useful, active, and dual-purpose, attracting many followers over the years. Their eggs range from lightly tinted to dark brown. Frank Reese of Good Shepherd Poultry Ranch

in Lindsborg, Kansas considers it "the perfect bird for outdoor production," along with New Hampshires.

Cochins are big, puffy chickens, with masses of soft feathers that create a rounded silhouette. Their feathers make them look even larger than they are underneath all the fluff. Calm and friendly, their disposition makes them excellent backyard birds. The hens are often good, broody hens and mothers. Cochins are a dual-purpose breed, big for meat and good egg layers. Most often, they are shown as exhibition birds.

Cornish take their name from the Cornish coast in England. Legs planted wide apart, the Cornish is a bulldog among chickens—a roast chicken on legs. Their heads are strong, with a small pea comb and wattles. They hold their short, hard feathers close, bringing out their vibrant colors and showing off their muscular physique. Keeping these burly chickens vigorous can be a challenge. They are inclined to gain weight, which is the meat producer's goal, but not any healthier for chickens than for people. They do well on pastures, where they can eat plenty of grass to keep their legs and feet bright yellow. Their natural inclination to develop muscle can also put on fat, which

SINGLE-COMB WHITE LEGHORN COCKEREL
Winner of first prize at Madison Square, New York, 1904. Owned and bred by E. G. Wyckoff, New York

The Poultry Book
Single Comb White Leghorn Cockerel (image courtesy of Cornell University Library)

interferes with fertility and egg production: a fat hen lays fewer eggs. Cornish need exercise as well as a nutritious, but not high calorie, diet to stay at their best.

Leghorns, with their yellow skin and prolific white eggs, originated in Italy and became popular everywhere they came to roost, though breeders focused on different qualities. In America, the Leghorn became "America's Business Hen" in the 1880s, setting it on the path to industrialization. Today, Leghorns have the most efficient feed-to-egg conversion ratio of all the Standard breeds. They produce more eggs in relation to the amount of feed they consume than any other breed. Standard quality Leghorns lay about two hundred and fifty eggs per year, for seven years.

Polish chickens aren't necessarily from Poland, although some of these popular chickens were undoubtedly raised there. The origin of this

Ideal Buff Cochins of 1895—By SEWELL.

Reliable Poultry Journal *Ideal Buff Cochins of 1895*

breed is lost in the mists of history. Its distinguishing feature, the crest—sometimes called a topknot or top hat—is full and round. Polish may have no comb at all, or only a humble one covered by the crest feathers. These chickens have remained popular through the centuries as good layers of white eggs. Today, Polish chickens are raised in seven color varieties, both bearded and non-bearded. A beard is the cluster of feathers on the throat, under the beak. Muffs are the feathers on the sides,

joining the beard to cover their faces from the eyes down to the throat.

Old English Games are the classic chickens, those of the English countryside as well as early America. They are homestead fowl, good layers and tasty meat birds who can find their own food and take care of themselves. In the United States, Old English Games were the choice utility small farm chickens into the early 20th century. ○

Stress and Surrender

Amphibious adaptation,

weightlessness,

toxicity and regeneration

Leilah Clarke

FLOATING FARM

It's been said that drinking seawater will make you mad, but you'd certainly be insane for thinking you could survive off salt water alone. The earthly plants and animals we consume also have this same requirement for fresh, salt-free water. Another seeming madness is that less than three percent of water on this planet is freshwater: water that isn't oversaturated with salts, and therefore can be used to drink or water plants.[1] It's a marvel we get enough for ourselves, let alone the crops we rely on for food.

Freshwater is a precious resource. Around the world, we pump it from source to farmlands and almost everywhere else to grow crop plants that keep the human population fed as we continue to multiply year-over-year.[2] As the global population rises, the earth's sea level does as well—more quickly than the population—as one of many results of climate change. It's predicted that sea levels could rise by as much as a foot by 2050, while demand for land to grow crops and build housing surges.[3]

Suitable farmland, too, is at risk. Farmland that is still above sea level over the next thirty years is predicted to reduce in quality as a result of a combination of nutrients being leached from the soil by unsustainable farming practices, meaning crop yields will fall and

what is produced likely won't be as tasty, either. With all of these factors coming to a head in an uncomfortably short amount of time, food production needs to change drastically, today.

I've long wondered how we will grow enough food in a world sinking beneath the salty sea, and where. A few years ago, I asked myself: "Why not grow plants on the sea?"

As the sea takes over the land, it seems fitting to use the water's surface for something traditionally land-based. Looking to nature, which so often has the answers we seek, I observed the water cycle in which our sun heats the ocean's water, causing evaporation into water vapor, collecting and forming clouds which eventually rain back down on the earth again, showering freshwater down onto the hills, forests, and farms—as well as back to the seas—providing plants and all other forms of life that reside here with that most precious resource we all rely on. I continued to search for answers until I came up with my solution and built a prototype.

On Floating Farm, plants are grown on a vessel floating on the surface of the sea. Plants are housed beneath a domed roof, under which water vapor is trapped, condensed, and delivered to roots when moisture evaporates by heat from the sun's rays. The hydroponically

Leilah Clarke *Floating Farm*

up to three people at a time able to sit on the exterior ring and reach in to plant, inspect, or harvest after removing a section of the dome. I initially imagined Floating Farm being used by communities in coastal areas but it works equally well in rivers, lakes, marinas, or any other bodies of water.

As with all great solutions, there are still environmental constraints that make Floating Farm less effective than desired. For instance, I carried out experiments over two years using prototypes to grow radishes, beets, and cress in Barbados and the United Kingdom, which revealed temperatures below 14°C failed to produce adequate water for plants, resulting in poor harvests. This means it probably wouldn't be suitable for cooler climates such as in the UK, though it still works well in the warmer summer months and can be relocated to different regions during cooler times of the year. Places with warmer climates would be able to benefit from Floating Farm year-round.

It's clear that the solutions Floating Farm can provide massively outshine the constraints. After natural disasters, Floating Farm can also help stabilize food security by being implemented in areas where farm lands have been flooded. There are a number of food crops that can be grown from seed to harvest in Floating

grown plants are provided with all the water they need to thrive by the sun's energy alone. From a bird's-eye view, Floating Farm consists of two rings encompassing another. The exterior ring provides the necessary buoyancy with the movement between the two parts allowing the plants in the interior ring to remain oriented in wavy conditions. It is designed to be tended by hand, with space for

Robert Dash *Lettuce Seeds*

Farm in just a month or two—spinach, radishes, beets, carrots, lettuce, microgreens, spring onions, snow peas, bok choy—making it appropriate to install them as soon as flooding has occurred to provide communities with food while water subsides and the lands heal and become fit for use again. In areas prone to flooding, Floating Farm provides a unique solution to growing crops, as it also offers a much greater diversity of crop production other than the typical soy and corn of which there are varieties bred for growth in floodplains.

Whether the flooding is caused by over-flowing riverbanks, monsoons, or freakishly large sea waves that find their way inland, Floating Farm would greatly help stabilize food security in regions hit hard by these destructive natural hazards which are increasingly severe and frequent as yet another result of climate change.

Plant crops are only just the beginning of the benefits that this invention can provide the world. With a little adaptation or clever anchoring, Floating Farm could easily provide additional uses: adding oyster nets from its underbelly, submerged in the sea, could provide oysters (and pearls, eventually) while simultaneously cleaning the sea by filtering out pollutants, excess nitrogen, and phosphorus. Acting as a buoy, Floating Farm could even be used to harness energy by anchoring them

to linear generators to convert the energy from the up-and-down movement of sea waves and tides to electricity.

Only by implementing new ideas and concepts will humanity be able to thrive and enjoy life on this planet. Things need to change for the better. They need to change now. We have the solutions. We just need the will. ○

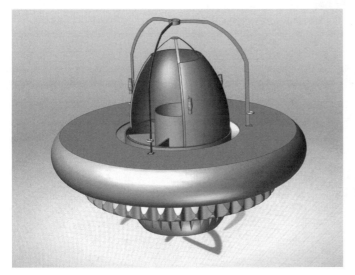

Leilah Clarke   *Floating Farm Schematic*

Notes

1. Lene Petersen, Martin Heynen, and Francesca Pellicciotti, "Freshwater Resources: Past, Present, Future," in Douglas Richardson, Noel Castree, Michael F. Goodchild, Audrey Lynn Kobayashi, Weidong Liu, and Richard A. Marston, eds., *International Encyclopedia of Geography: People, the Earth, Environment and Technology* (Hoboken: John Wiley & Sons, 2017): 1–12.

2. Mengistu M. Maja and Samuel F. Ayano, "The Impact of Population Growth on Natural Resources and Farmers' Capacity to Adapt to Climate Change in Low-Income Countries," *Earth Systems and Environment* 5, no. 2 (June 2021): 271–283.

3. Rebecca K.Priestley, Zoë Heine, and Taciano L. Milfont. "Public Understanding of Climate Change-Related Sea-Level Rise," *PLoS ONE* 16, no. 7 (2021).

4. David Tilman, Christian Balzer, Jason Hill, and Belinda L. Befort. "Global Food Demand and the Sustainable Intensification of Agriculture," *Proceedings of the National Academy of Sciences* 108, no. 50 (February 2011): 2026–2064.

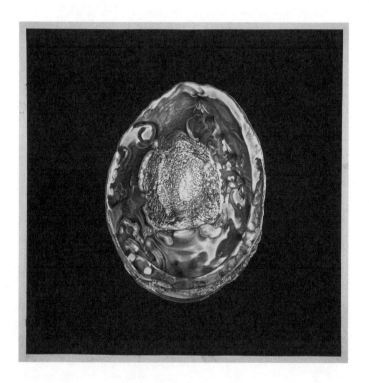

Tanja Geis *Largest Abalone Ever Found*

Tanja Geis   *Purple Sea Urchin*

Austin Miles

PERMANENT TOXICITY

Around the world, mining operations have blasted land into alien landscapes. The violence of these operations lingers on long after the initial literal blasting of mines—ecological devastation like mining introduces new modes of becoming shaped by violence. In the lands surrounding my hometown in Appalachian Ohio, abandoned mines pervade, and as a result, a toxic seepage called acid mine drainage (AMD) constantly leaks into the region's streams.

AMD is one link in a concatenation that begins with extractive capitalism and subsequent neglect, whose end result is, to varying degrees, permanently disturbed ecosystems. It forms after a mine is abandoned, once its remaining pyrite-rich coal is exposed to water and air successively, producing the sulfuric acid characteristic of AMD. Once begun, AMD's seepage can last hundreds or thousands of years.

AMD is prevalent around the world. In the United States, an estimated five hundred thousand abandoned mines exist in thirty-two states.[1] AMD formed in these mines has impacted over 16,900 kilometers of streams in the US alone.[2] In Appalachia, there are an estimated three to five thousand abandoned mines impacting seven to thirteen thousand kilometers of streams.[3]

AMD's ubiquity, persistence, and insidiousness on ecosystems requires that they be restored, a practice that is typically the domain of restoration ecologists. In their book *Making Nature Whole*, botanist William R. Jordan III

Jin Zhu *Boron*

and historian George M. Lubick characterize restoration ecology as "the attempt to reverse the effects of novel influences on preexisting ecosystems" as a means to ensure the survival of ecosystems in a natural state. Restoration is an endeavor Jordan and Lubick call an ecosystem's "rescue, so to speak, from history."[4]

Restoration, as the authors put it, is "a response to the mixture of regret, nostalgia, and curiosity some feel on looking back at the 'original' landscape they and their predecessors have altered, often beyond recognition."[5] Restoration, evidently, is as much about recovering Eden as it is about remediating an anthropogenic disturbance like toxicity.

In his critique of restoration ecology, Jeremy Trombley writes, "By relegating spatial ruptures (e.g., wilderness) to the past (e.g., as "Edenic narrative"), nature is made to seem transcendent—outside the scope of human history." Restoring ecosystems is as much a restoration of this rupture as it is of biotic communities. The dream of restoration, according to Trombley, rests on the faith of the inevitable march of progress. He writes that restoration only requires "a continuation of modernist progress toward a sustainable future in which human impacts on nature are reduced so that nature can return to its undisturbed state"—a rescue from history that must be continuously engineered and managed.[6]

Ecological restoration typically engenders a transformation, one of nonhuman nature into an image that powerful people employ to shape restoration into what they think nature should be. But in the blasted landscapes of Appalachian Ohio, restoration mutates. Ordinarily, to be considered successful, ecological restoration must produce self-sustaining, resilient ecosystems that require minimal human intervention. In Ohio's abandoned mine lands, however, the constant flow of AMD necessitates permanent additions to restore stream ecologies.

Among the possible additions that restoration practitioners use, the doser is a particularly striking example of a stream ecosystem's

Austin Miles
Opaque waters of Sunday Creek, impaired by acid mine drainage

permanent modification. A doser is essentially a silo that provides a constant input of an alkalinizing material, like lime, to counter the acidity of the AMD. The alkalinity raises the pH of the water, which removes the dissolved metals. They turn into solids that coat the stream bottom for a stretch downstream of the doser, an area often called the "sacrifice zone" because the metals make it uninhabitable.[7] If restoration is successful, the water chemistry below the sacrifice zone is similar to that of one unimpaired by AMD.[8] Restored water quality doesn't guarantee the return of annihilated insects and fish, but it does provide an opportunity for their return. It's all anyone can do for these streams, feasibly.

The stream's health as an ecosystem is inextricably tangled with the doser—it's less of a discrete stream interacting with a discrete doser and more of a doser stream. In *How Like a Leaf*, the science studies scholar Donna Haraway describes "naturecultures" as "implosions of the discursive realms of nature and culture."[9] Like the monolithic categories "nature" and "culture," the stream and doser as they interact in streams in Appalachian Ohio have also imploded as discursive realms—it's impossible to consider them apart.

The doserstream is a kind of technoecosystem. The ecologist Eugene Odum considered technoecosystems to be "fuel-powered" realms created by "urban-industrial" society. He contrasted the technoecosystem with the solar-powered "natural life-support ecosystems" and hoped the natural and technoecosystems could "co-evolve for mutualistic co-existence."[10]

I have a different vision, one that aligns with the Out of the Woods Collective's cyborg ecological approach. They write that cyborg ecology is premised on a "suspicion of 'organic holism'—the notion of an organic wholeness—

and sharp binaries between natural and artificial, living and nonliving."[11] Cyborg ecology, like Haraway's naturecultures, entails imploding the categories—like human and nature—organizing the world according to a violent and domineering logic.

The doserstream is a gross subversion of the "organic holism" of nature. As a kind of technoecosystem, it points not towards mutualistic coevolution of separate spheres comprising the artificial and the natural, but grotesque intertwining—and it is grotesque. The dosers are large and metal—one I've visited had piles of lime scattered about a concrete platform above orange water. A clank emanated metronomically from its interior, followed every time by a flushing of milky water out its effluent pipe. Downstream of the pipe the stream is a weird mix of orange iron sediment and lime. It's not utopia—but for the stream dwellers, it makes things livable. The promise of technoecosystems like the doserstream, then, is to make Earth livable after ecological devastation. In her article in *Logic*, Alyssa Battistoni asks at one point, "Can human technologies or human labor substitute for the nonhuman work done by other organisms? Or are there certain kinds of work that only nature can do?"[12] In the wake of catastrophes like climate change, new niches will open up that only tech can fill. ○

Jin Zhu *Glen Canyon*

Notes

1. Dan Peplow, "Environmental Impacts of Hardrock Mining in Eastern Washington," (Seattle: The Water Center, 1999).

2. Alan T. Herlihy, Phillip R. Kaufmann, Mark E. Mitch, and Douglas D. Brown, "Regional Estimates of Acid Mine Drainage Impact on Streams in the Mid-Atlantic and Southeastern United States," *Water, Air, and Soil Pollution* 50, no. 1 (March 1990): 91–107.

3. B.H. Hill, J.M. Lazorchak, A.T. Herlihy, F.H. McCormick, M.B. Griffith, A. Liu, P. Haggerty, B. Rosenbaum, and D.J. Klemm, "The Extent Of Mine Drainage Into Streams Of The Central Appalachian And Rocky Mountain Regions," (Paper presentation, The EMAP Symposium on Western Ecological Systems, San Francisco, CA, April 6–8, 1999).

4. William R. Jordan III and George M. Lubick, *Making Nature Whole: A History of Ecological Restoration*, (Washington, DC: Island Press, 2011): 3.

5. Jordan III, 15.

6. Jeremy Trombley, "Watershed Encounters," *Environmental Humanities* 10, no. 1 (May 2018): 107–128.

7. Natalie A. Kruse, Lisa DeRose, Rebekah Korenowsky, Jennifer R. Bowman, Dina Lopez, Kelly Johnson, and Edward Rankin, "The Role of Remediation, Natural Alkalinity Sources and Physical Stream Parameters in Stream Recovery," *Journal of Environmental Management* 128 (July 2018): 1000–1011.

8. Edwin E. Herricks, "Recovery of Streams from Chronic Pollution Stress—Acid Mine Drainage," in J. Cairns Jr., K.L.

Dickson and E.E. Herricks, eds., *Recovery and Restoration of Damaged Ecosystems* (Charlottesville: University Press of Virginia, 1977): 43–71.

9. Donna J. Haraway, *How Like a Leaf: An Interview with Thyrza Nichols Goodeve*, (New York: Routledge, 2000): 115.

10. Eugene P. Odum, "The 'Techno-Ecosystem,'" *Bulletin of the Ecological Society of America* 82, no.2 (April 2001): 137–138.

11. Out of the Woods Collective, "Contemporary Agriculture: Climate, Capital, and Cyborg Ecology," *Libcom.org* (July 2015).

12. Alyssa Battistoni, "A Repair Manual for Spaceship Earth," *Logic*, December 7, 2019.

Suzanne Husky *Let Beavers Do the Job*
Immediately after the fires, the government decides to collaborate with the beaver people

Madeleine Granath

ON FARMING

Relinquishing control, a lifestyle.

It is accepting that you are neither the steward of life or the usher of death.

Life wants to live, and so it will.

We are only channeling what flows like veins from its source.

The sun, a ripened apricot caught in mesquite branches.

A dozen chicken eggs in coat pockets.

In my mouth, raw cauliflower dipped in honey.

Like water in our hands, life runs on.

Melina Roise

DIS-ABILITY AND LAND LOVE

An ongoing romance between a small tired body and the small tired land
I most days tend

At the start of this season, I sprinkled extra poppy seeds inside the freshly weeded garden box on my front stoop. I despise wasting seeds, or even sitting on them until next spring, so I often end up sprinkling extra somewhere I believe they may grow. One of my early gardening teachers taught me this. She told me she gardened "with a lick and a prayer."

I did not have the energy to put much into their livelihood, but I did indeed pray, spend-

↑ → Madelaine Corbin (photography by Taylor Kallio and Prairie Ronde Artist Residency) *A Moon in a Meadow in a Moon*

ing more time showering words of affirmation than fertilizer. I continued to sleepily sprinkle water onto their new home, and, eventually, they sleepily rose from the dirt—slowly—and sprouted a few leaves—slowly. While witnessing their lack of care towards my impatience, I noticed myself jealous of their slow-moving possibilities.

As are many of us raised under colonial human-land-animal divisions, I have been practicing and reading and thinking about ways to reform my body-land relationship. This relationship is strained by a mediator—

a diagnosis of mine, which of no fault of the soil makes our days together more tense—as I am tired.

I have been upset at the land I tend for its constant need for attention. It feels, when I wake on the verge of being too tired to get dressed, like a personal fight. This season, I'm experimenting in reimagining my relationship with the soil patches I exhaust myself spending time with, continually recommitting to a happy ending to the love story between them and my dirt stained hands. As a person with postviral chronic fatigue (otherwise known as Myalgic encephalomyelitis/chronic fatigue syndrome, or ME/CFS), a syndrome very similar to long COVID-19, there are days when I question why I would ever voluntarily put myself through such long hours and laborious tasks.

Indeed, two years ago I was completely unable to, as many of those with a similar diagnosis are. My friends and colleagues with no such diagnoses still reflect on the Thursday evening feeling, *this career may kill me.* And, even so, we all feel unable to do anything else. By no means am I alone in such feelings of possible masochism: an estimation by the US Department of Agriculture Economic Research Service found that about nineteen percent of farmers have a disability that impacts their ability to work.[1] And if mental health struggles were counted here as a disability, this number would be much higher.

I wake up each day fearing that my body will simply not work. Some days, this happens. I am okay for a while, then my ears start to ring and my eyes get heavy. My brain goes foggy and the sentences I speak become almost incoherent—a brain fog interrupting the wiring between my brain and mouth. Once in a while, I get dizzy, weak, and incapable. Any good relationship therapist may question whether the root of the problem is not in itself the trigger, but the reaction. There is no direct correlation between my hands in the soil and my continuous exhaustion. It is no fault of the land that we are trying to control it for production. It is no fault of the chickweed seed for simply trying to survive. It is no fault of my body, or yours, that we are tired.

Sometimes experimentation and resulting solutions are not in themselves tangible. Sometimes our problems exist without solution, while those of us with chronic diagnoses are well-versed in healing as an impossibility. The solidarity of knowing the soil is indeed tired too, stretched thin in giving, and likely bracing itself for another season, our relationship to the land can mend in reimagined reciprocity that includes compassion for more than just a few sprouts of poppy. My poppies are blooming now, after all, simply in response to my prayer.

In honor of my poppies and dirt-stained hands, alive and happy, I have decided that rather than allow spite to grow inside of me at how *oh my goodness, it cannot be possible that the flowers need deadheading yet again*, I have adapted a new maybe-imaginative-possibly-true relationship between my body and the land. Instead of a straight line from my sore arms into the ground, what if these transfers of

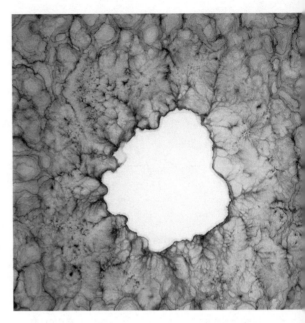

Toni Gentilli *Cavity*

work and energy resemble more closely an ever-entangled web, where energy flows both and every way, from myself to the soil, to the soil to myself, and when no separation exists (for in fact I am often covered in soil), in spirals of here-and-there?

Of course, imaginative reframing of vital movement has little, if any, impact on our actual energy levels. But a bit of imaginative play, a bit of adaptation to a problem with no solution, may just revive us through the last root harvest. ○

Note

1. Christina D. Miller and Robert A. Aherin, "The Prevalence of Disabilities in the US Farm Population," *Journal of Agricultural Safety and Health* 24, no. 4 (January 2018): 243.

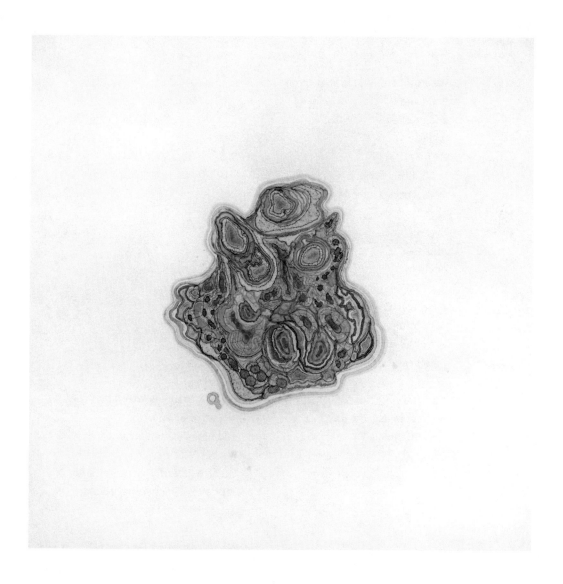

Toni Gentilli *Mass*

Phil Cordelli

BETA VULGARIS

We crowd around the shaven pile
child and I
and mother
and eat

in handfuls burst
folds where stitched

Big flag
pink flag
or chard

turning in the cavalcade
of morning, in the cadence
of morning, draped in cakeswirl,
mercilessly stunning

The skin slips right off
we have to leak the rocks
they leak to lead

it's easy to forget we're a planet
we turn and return
and now the light bending

Green gobble of swirl we weather
little swirls connect at the edges
with other little swirls

quick on the cast iron flakes
sputter alongside the water in the pan

H.e. Haugenes

REIMAGINING ANIMAL SANCTUARIES

I thought vegan animal sanctuaries were a beneficial response to factory-farmed animal abuses until I interned at one. Throughout my experience, I came to believe that imposing human values like longevity and security through pharmaceuticals and fences, rather than reintroducing animals to some semblance of ecological harmony, is savioristic zookeeping. If that strikes a cord with you, I hope you will keep reading.[1]

I've been vegan for the past seven years for a mix of environmental, moral, and spiritual reasons. Throughout this time, I have been both intimidated and intrigued by the work of animal liberation vegans, so I thought I would give it a try interning at a well-known vegan animal sanctuary this past fall. Vegan animal sanctuaries are generally run by non-profit organizations. They typically rescue factory-farmed animals, like cows, sheep, chickens, horses, alpacas, and goats, and raise them outside of the factory setting while also promoting a vegan lifestyle. Living on site, I hoped to be immersed in this counterculture and learn how to raise animals outside of the context of consumption. Yet the experience left me shaken. It didn't feel like a sacred place of refuge and it shifted my idea of what previously farmed animals deserve in their life post-captivity.

Sandra Taggart *Sheep*

More than anything else, the place reminded me of a zoo: overgrazed enclosures of animals separated by species, caged-in waterfowl with no more than a shared fifty-gallon trough to swim in, huge amounts of hay, straw, and food

153

Shauna Lee Lange   *The Stable*

opportunities for symbiosis. Many animals were medicated past the point of their natural lifespans. Feed and straw were purchased en masse and shipped in, while the earth, which could be managed to provide these resources, was not cared for. Manure was not returned to the soil. I was stunned when I was told by staff that it was considered exploitative to use manure for soil health by the wider animal liberation community, and that roughly 1,500 cubic feet of manure and soiled straw were dumped in an unmanaged pile at the sanctuary weekly as standard practice. This did not come just from sheds and coops, but also from the pastures.

This struck me as a grave disrespect to the animals. Animal manure serves a function to the soil, but I found that at the sanctuary, having an animal serve any function, even an ecological one, was considered exploitation. To me, rejecting the benefits of manure was an offense to both the animal and the earth. Humans have already removed themselves so much from the natural world; to do the same to animals is not freedom. It is instead an anthropomorphizing imposition.

There is still room for us to reconcile the broken way in which animal sanctuaries using the same staunch vegan model function today, through an intentionally ecologically minded model. For this to happen, those who run sanctuaries would have to first and foremost honor relationships.

The Relationship Between Different Types of Animals

Rather than separating animals by species, animals could commingle or graze the same pasture at different times, and reap the benefits of being in relationship. For example, pastures pillaged by pigs could be rested under the lighter

trucked in, and huge amounts of shit scooped out. Unlike a zoo, this sanctuary doesn't profit off the animals as entertainment; however, they do get funding by marketing cute pictures on social media.

Most notably, I was overwhelmed by the ecological disconnection. I studied ecology, and am constantly awed by the ways that all creatures in nature play a role in creating harmony. However, there, different species did not share space and thus there were no feet of chickens. Meanwhile, the birds could eat maggots from the animal dung, limiting disease. Grazing cattle and goats in the same pasture can help manage different types of foliage, keeping the vegetative life balanced as well. Numerous other benefits come from agro-ecology, and it would mean fewer fences, too.

Between Animals and the Land

Rather than acquiring as many animals as possible, prioritizing space for each would prevent overcrowding, which compromises wellness. Some of the practices used by regenerative livestock farmers could be explored, namely rotational grazing. Through rotational grazing, cows and other animals can continually have fresh food to forage, rather than buying so much food. The earth can sequester carbon, biodiversity is bolstered, and disease is lowered. Drought tolerance is heightened as water absorption increases, soil compaction of pasture is diminished, and the entire ecosystem is given a chance to thrive.

Between Manure and Soil Microbes

Chicken coops and animal sheds need to be cleaned, but manure should not need to be picked out of pastures. It should integrate into them, providing microbial abundance and nourishing the earth. With the implementation of rotational grazing, this is easy, healthy, and triply impactful in helping pasture thrive.

Between the Living and the Dead

It would mean honoring the death of each animal not as an end, but as a return to the earth. Each time a singular being passes on, the material that made up their physical form nurtures future life. Imposing the human obsession with longevity on animals has created a norm of dependence on pharmaceu-ticals for medication, bolstering an already corrupt industry. Why is a long life in captivity better than a short life in the wild?

Between Humans

This means ensuring a manageable workload of animal care among employees to keep morale high and ensure humans are being cared for, too. Human boundaries can be neglected in the vegan world, but communicating and honoring them is powerful.

I think that farms which try to imagine animal liberation can be a powerful and beautiful thing, and we need to reimagine what that means. Anthropomorphization and saviorism, neither frees animals nor is in harmony with nature. With collaboration and open minds, I know spaces of sanctuary are possible. ○

Note

1. Before I go on, I do want to acknowl-edge that I only experienced one vegan animal sanctuary non-profit, for a quite limited time. However, this organization is linked to a wider network of similar organizations and from my research and the stories I have heard from others, they seem to often share similar beliefs and models.

Brett F. Braley-Palko

THE BARN STAYS EMPTY

My days usually begin at about eight in the morning. This, of course, is something that most other farmers may envy. I do not have livestock to care for. Instead, I let the warm bodies of my dogs nudge up against me. They're all old now, so they like to sleep in. It's an affliction that I share, too, I'm sorry to say.

It wasn't always this way. I once had chickens—lots of them. A veritable flock, even. But three Decembers ago, I lost them all. The culprit was a weasel, the vicious little thing. It used the old tunnels from long-gone rats to crawl into the barn, up through the ground, and bite the necks of thirty-four hens. We had one survivor. She died a week later from shock, I think. She was a bantam—how much could her small heart take?

I wonder the same about myself. How much could mine?

It wasn't until the chickens were gone— unceremoniously tossed into a large moving box until my husband came home to sort the mess out with me—that I realized how much of my life revolved around their little habits. Six in the morning, the alarm went off, and I would be groggy in my hellos to their antici- patory chatter.

I would throw the barn doors open and they would fight their way to the front of the line for the best morning worms. In the winter I would wake up even earlier to bring them warm oatmeal and a kettle of boiling water to thaw out what had frozen overnight. Anything to make their little lives happy. They did nothing but get fat and keep me company.

Now, my mornings are open. I can sleep in and not feel guilty. No rushing to the barn to hear the aggravated, accusatory clucking from my girls. They say that chickens have their own language; I believe most of what they say is just one complaint after another.

It is strange how much of my day was spent thinking of them. When I made dinner, I would keep back the vegetable scraps for their breakfast the next morning. The hours I spent worrying when a hen was sick. The money! The vet trips! The nausea of performing small backyard surgeries on bumblefoot and sour crop.[1] The elation of holding a still-warm egg in my palm, the subsequent nips on my hand from a broody hen.

I miss sweet moments the most, the kind that I never thought to photograph because I seemed to have always taken my time with the hens for granted. There was a bantam who sat on my shoulder while I cleaned the coops. I would feed her berries and she would fall asleep in the pocket of my old Barbour jacket.

Anselmus Boëtius de Boodt *Weasel* (Mustela nivalis)

It's hard to say if I begrudge the weasel. Chickens, I must say, are an easy enough target. Mine especially, who were slow and mammalian and, as I mentioned, fat. I blame myself, of course. I didn't think to seal the tunnels where the rats had come in. I didn't think to put their roosts higher. But would it have made a difference? We had five good years. The weasel had one good meal. Was it a trade-off? Perhaps. I never set a trap to catch the damn thing. I just hoped it wouldn't happen again.

And now I sleep in until eight in the morning. We don't own hens anymore. I'm too scared to go through it all again. I think about ducks and quail, turkeys, perhaps. Something sturdier. Something unpalatable to a weasel or a fox or some such other boogie man in the night. Something that won't leave me so empty once they're gone. ○

She was a show chicken I got on Craigslist. She was used to being handled and was too old when we got her to lay more than one egg a month. She died in the massacre along with her sisters. I hope she knew she was my favorite and I hope she did not suffer.

Note

1. Bumblefoot is a treatable foot abscess caused by an infection from a scrape or cut that is then impacted with dirt or other irritants. Sour crop is a yeast infection in the crop, a small storage sac that is part of the digestive system in chickens.

Over the Course of the Day

Vegetative strategies,

dry farms,

bioplasticity

Don Tipping *Garlic*

Don Tipping

NEW DRY FARMING STRATEGIES FOR THE PACIFIC NORTHWEST

As a farmer of nearly thirty years, I am intrigued by the various ideas around adapting agriculture to my bioregion and its changing climate. I've conducted a wide variety of experiments over the decades: planting traditional vegetable crops with a variety of plant spacings, different mulches, no-till strategies, and nonirrigated dryland approaches. The primary data and subsequent conclusions that I have come away with are that the common crop species that many of us may grow have traveled a long pathway, from wild crop relative to the domesticated, refined variety that exists today. As agriculture evolved, these common crop species became more sophisticated, with more careful planting, spacing, fertilization, irrigation, trellising, and mulching. In our modern approach to agriculture, we have generally tried to optimize growing conditions as much as possible. Historically, much of early agriculture was dryland due to the lack of irrigation techniques and technology. Decreased rainfall patterns over much of the West have compelled many growers to explore alternatives to modern irrigation strategies.

The definition of dry farming, as explained by Lucas Nerbert of Oregon State University (OSU) is farming without irrigation during an arid growing season, by accessing soil moisture that is replenished by the wet season. Some distinguish 'dry farming' from 'dryland farming' by setting a threshold precipitation level: say, twenty inches or more annually is dry farming, and less than twenty inches annually is dryland farming.

This borrows heavily from Mediterranean dry farming traditions and, in practice, works particularly well in certain areas of the maritime Pacific Northwest. The Dry Farming Collaborative (DFC) was founded in 2016 by a group of growers, educators, researchers, and plant breeders to increase awareness of practices and strategies that support crop production with little to no irrigation.[1] As a convention within the DFC, to participate in a dry farmed trial, you are only to irrigate during planting—if there is insufficient rain to establish your plants, for example—and then no irrigation at all after that. Having an established dry farming protocol has been useful to determine important site factors and drought-hardy cultivars that increase dry farming success.

However, as I have listened to more and more people working with water conservation in their own areas across the western United States, I feel less inclined to attach an exact definition to dry farming. Growers have been doing incredible work, some for many generations, using minimal water and in climates that are inherently not very suitable for dry farming as has been defined in our DFC variety trials.

Amy Garret, another educator at OSU, helped to initiate the DFC, which the OSU Dry Farming Project supports by facilitating communication, creating space for information sharing, coordinating participatory research, and developing resources to assist growers new to dry farming.

In learning how to work with the land while aspiring towards a reverential and regenerative approach, I have come to realize that much of the predominant settler mindset is still in a fairly adolescent colonizer, or "pioneer" mindset. Consider that Oregon did not achieve statehood until 1858 and the first settlers arrived in the 1820s.[2] Further, due to the abundance of native edible plant and animal resources, the Indigenous peoples of this area—Takelma, Karuk, and the Dakubetede— never developed what we may recognize as agriculture, but their land stewardship was arguably far more sophisticated and regenerative.[3] Many of us are still figuring out what and how to grow here. To demonstrate this point, in the Pacific Northwest, I have had successful harvests from Russian short-season pomegranates, Arbequina olives, peanuts, and pineapple guavas.

Today's farmers will all be well served to continue to experiment and refine both the techniques and the cultivars that we marry towards successful growing in what appears to be an increasingly arid West. So much of this work is site-specific and is dependent upon the water table, moisture holding capacity, soil organic matter levels, aspect, rainfall distribution, choice of crops, desired yields, and so on. Experimenting with various no-till and dry farming approaches for growing crops such as corn, beans, squash, tomatoes, and melons, I may have achieved success in growing, but the yields were always significantly less than growing the same crops under ideal circumstances and conditions.

Many of the summer garden favorites originated in the humid subtropical region of Mesoamerica. As these crops traveled north through exchange and trade, they underwent centuries of adaptation to local conditions in order to successfully produce seed and continue radiating out from their center of origin. An international team of scientists led by Dolores Piperno, an archaeobotanist at the Smithsonian's National Museum of Natural History, and Anthony Ranere, professor of anthropology at Temple University in Philadelphia, discovered the first direct evidence that indicates maize was domesticated 8,700 years ago, the earliest date recorded for the crop.[4] However, it is thought that corn did not arrive in what is now the United States until one to four thousand years ago, largely because the ancestor of modern corn is a photo-sensitive grass called Teosinte *(Zea mexicana)*. This grass doesn't initiate flowering and seed formation until the day length begins to decrease after the summer solstice, while day length is fairly consistent from summer to winter in the tropics. When Teosinte is grown at our farm—latitude 42° north—it grows luxuriantly tall leaves, up to nine feet, but does not initiate seed formation until September. We typically get our first fall frost in October, thereby ruling out a successful seed crop.

It took many hundreds of generations of corn to acclimatize to the long days of more northern latitudes. The ancestral corn from one thousand years ago that Apache peoples

H.H. Iltis and Doebley   *Taxón:* Zea mays subsp. mexicana
(image courtesy of The Trustees of Indiana University)

A

B

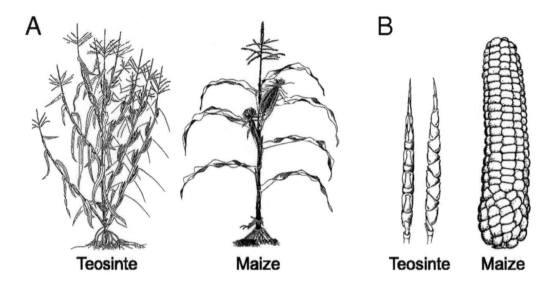

Teosinte **Maize** **Teosinte** **Maize**

Proceedings of the National Academy of Sciences
Teosinte and Maize (image courtesy of PNAS)

adapted to summer monsoon rains to irrigate.[5] This can be seen in the deep taproot that traditional Southwest corn varieties develop quickly upon germination. Contrast this to modern sweet corn that develops a multi-branched, fibrous root system upon germination, having adapted to ample irrigation and fertilizer.

Modern agriculture is constantly aiming to optimize for a given crop, and corn is certainly no exception. Many growers will transplant sweet corn to ensure more successful production, even though it has been planted directly in the soil the world over for centuries. This is leading the plant in the direction of adaptation to this agronomic technique.

I feel that today's land stewards must optimize our use of the land to honor the supreme blessing of having access to it while honoring that we appear to be in the early stages of what may be a prolonged drought cycle. Storing and redistributing water is a concurrent approach that stewards would be wise to employ. Growing heat-loving summer crops can be done quite successfully if one is confident in their access to the necessary water through the heat of late summer. On my home farm here at Seven Seeds Farm, I have constructed a total of nine ponds to capture and store rainwater on my farm, to have it available to finish out summer crops.

My own experiments here on the farm with no-till and dryland strategies turned out to be fairly labor-intensive. I have come to the conclusion that as much as I am fascinated with novel permaculture approaches and how they seem to offer the promise of saving labor and moving towards a more self-regulating ecosystem, they almost universally result in significantly lower yields. The biodiversity and metastability on our farm is much higher than a monocrop. Some of the strategies that I have tried include no-till with mulch, no-till with cover crop scythed down for mulch and then transplanting into the mulch, unirrigated Three Sisters plantings of corn, beans and squash, unirrigated garlic, and wider spacings of squash grown without irrigation. In all of these experiments I felt as though I was doing the right thing and was hopeful for the outcomes; nevertheless, I saw low yields, small fruits, or complete crop failures. The yields that I did obtain through experimentation came in the form of knowledge, experience, and wisdom.

Often permaculture is advertised as "no-work" gardening, popularized by cofounder Bill Mollison's classic image of the "designer recliner" where one simply kicks back and watches it all grow.[6] Maybe in the tropics this is possible, but I strongly doubt it. Considering the high value of farmland in western Oregon, the cost of deer fencing, labor, and the other inputs, I am doing my best to optimize yields, within limits of the soil productivity. I have taken the cue from permaculture design to focus more on water harvesting and storage to ensure ample water during dry spells.

My two cents is that most of the crops that we grow during the dry season have thousands of years of selection towards optimizing yields or are grown in regions with summer rainfall. Traditional Three Sisters plantings of corn, beans, and squash in the desert Southwest were planted ten to twelve inches deep in sandy soils in advance of summer monsoon rains. We don't have those rains here in Oregon, and if we planted seeds that deep they would most certainly rot in our cool, moist soil.

Over millennia, civilizations have redirected water in canals, aqueducts, and channels to create resilient agriculture by averting undesirable flooding, prolonging water availability, and increasing arable land. History bears this out, and a review of the civilizations that were able to feed large populations had much to do with their abilities at civil engineering and techniques for irrigation. The crops they worked with evolved and adapted to their environment over the generations due to a beautiful inevitability with seed saving, or grain harvesting in which the plants that perform the best and yield the most will always represent themselves in each succeeding generation as a larger percentage of the total population.

Conversely, the plants that do not yield as well—or those that perish outright from pests, disease, or other environmental stresses—will manifest as an increasingly smaller percentage of the total population. Even without thoughtful selection which would involve only saving seeds from the top 10 percent of the plants in a population, mass selection (the term to describe this approach of simply pulling one's planting stock for the next generation/season from the total harvest) trends towards adaptation to soil, climate and the total environment in an increasingly logarithmic fashion.[7]

The tumultuous rise and fall of civilizations throughout history mirror periods of plenty—and those of scarcity have often been due to the lack of availability of water to grow crops to feed the workers necessary to build and maintain extensive public works, temples, roads, fields, and mines. The civilizations that developed reliable systems for maximizing agricultural yields were able to placate the masses to continue to support the rulers. This was often carried out with a combination of irrigation canals, ditches, flood control structures, granaries for storage, roads for distribution, and crop improvement through selection and domestication of a diverse suite of plants. Dryland agricultural societies were limited in their spread and influence due to their agricultural output being inextricably linked to rainfall and how the rainy season overlayed upon the growth cycles of their key crops.

According to my research and also the conclusion of NASA/NOAA on the subject, most civilizations failed due to a changing climate, drought, and declining crop yields coupled with a top-heavy consolidation of people in managerial/political leadership positions. Seems like the industrial civilizations of the global north are in such a circumstance presently.[8]

An overly simplistic approach to dryland farming might become just another sustainable agriculture buzzword like no-till, regenerative, or local, lacking the depth of having been developed over many seasons of actual practice in a specific region. Where I live and farm here in Southwest Oregon is characteristic of the wider Pacific Northwest region. The climate alternates from a moist temperate climate for half the year and then a dry Mediterranean climate in the summer. Our rainfall season begins after the fall equinox and typically continues to the end of May, although some years May can be dry and warm, or June can be rainy and cool. Given our relatively mild winters, many crops can be planted in the fall, including garlic, onions, leeks, overwintering cereal grains, brassicas, chard, beets, radicchio, parsley, cilantro, and parsnips. Traditional herbs that are predominantly of Mediterranean descent that we grow for seed also include mints, rosemary, thyme, catnip, clary sage, rue, hyssop, and others. Our farm system has gradu-

ally evolved to become primarily focused upon seed production, as this most accurately mirrors the reproductive cycle of many of the significant vegetable and herb crops that we work with.

Consider a *Brassica oleracea* crop such as Lacinato kale that was domesticated in the Mediterranean Sea area of modern-day Italy, that originated as a wild plant whose seeds likely sprouted with the coming of early fall rains, becoming a 12–18" tall leafy green plant over winter, and then forming immature flower buds with the increasing day length and the warmer temperatures that February brings at that latitude. Then it would reach full flower in April, and, as the rains began to subside in May, seed pods form, swell, and naturally dry down towards brittleness in the heat of June and then finally shatter in late June or early July.

A central tenet of permaculture, or any whole systems approach (including regenerative farming, natural farming, and holistic resource management) to farming and land stewardship, is pattern recognition. So, considering the patterns of the Mediterranean climate that is common throughout the dry season in the Pacific Northwest, it seems to make sense to establish crops with the fall rains and the support systems that allow them to grow and flourish when the moisture is present in fall, winter, and spring, which would be beneficial at the onset of our fairly predictable dry season in May and June. In my estimation, this is what dryland agriculture looks like for Southwest Oregon and similar bioregions. Crops such as garlic, onions, fava beans, and many brassicas can be planted anytime between the fall equinox and Halloween (precisely when our fall rainy season begins), then only sparingly irrigated if May and June are drier than normal to produce admirable yields.

These ideas and techniques emerging from applying theory to practice spring from my deep curiosity in reviewing ancient histories of agrarian civilizations. Given the high value of farmland throughout the Pacific Northwest, today's land stewards must honor the supreme blessing of access to land by optimizing its use. I have always been inspired by the novel approach to problem solving as so wonderfully demonstrated by R. Buckminster Fuller who famously said, "You never change things by fighting the existing reality. To change something, build a new model that makes the existing model obsolete."[9] So in that spirit, may we all keep experimenting with new crops and novel ways to grow them in harmony with the land and her resources and not lose sight of obtaining a yield so we can share with others. ○

A version of this profile with the same title was initially published on Siskiyou Seeds blog, and was printed with permission.

Notes

1. Amy Garrett, *Intro to Dry Farming Organic Vegetables: Dry Farming in the Maritime Pacific Northwest* (Oregon State University: Oregon, 2019).

2. The OE Staff, "Colonizing the Oregon Country." Oregon Encyclopedia, 2020, *oregonencyclopedia.org/packets/4.*

3. US Fish and Wildlife Service, "Traditional Ecological Knowledge for Application by Service Scientists," February 11, 2011.

4. World Archaeology Staff, "Maize Domestication." World Archeology, May 6, 2009. *world-archaeology.com/world/ south-america/mexico/maize-domes- tication.*

5. Smithsonian, "Scientists Overhaul Corn Domestication Story With Multidis- ciplinary Analysis," December 13, 2018.

6. Cincinnati Permaculture Institute, "What is Permaculture?" Cincinnati Permaculture Institute. *cincinnatiper- macultureinstitute.org/read-me.*

7. John Navazio and Jared Zystro, "Introduction to On-farm Organic Plant Breeding," Organic Seed Alliance, 2014. *seedalliance.org/publications/introduc- tion-to-on-farm-organic-plant-breeding.*

8. Emily Sohn, "Climate change and the rise and fall of civilizations." NASA, January 20, 2012.

9. Daniel Quinn, Beyond Civilization: Humanity's Next Great Adventure (New York: Crown, 1999), 137.

Gary Snyder

SONG OF THE TASTE

Eating the living germs of grasses
Eating the ova of large birds

    the fleshy sweetness packed
    around the sperm of swaying trees

The muscles of the flanks and thighs of
        soft-voiced cows
    the bounce in the lamb's leap
    the swish in the ox's tail

Eating roots grown swoll
        inside the soil

Drawing on life of living
    clustered points of light spun
        out of space
hidden in the grape.

Eating each other's seed
        eating
    ah, each other.

Kissing the lover in the mouth of bread:
        lip to lip.

From Regarding Wave (New Directions, 1970)
Printed by permission of the author.

Supermrin, Jessica Fertonani Cooke and Natalia Mount

HOW TO GROW A FIELD?

FIELD is a bio-arts project and research laboratory creating a new discourse around our relationships with land and nature. FIELD develops land interventions, drawings, sculptures, and performances that subvert the aesthetics of control perpetuated within contemporary civic parks.

I met Mrin in 2018, when she proposed FIELD for Frank H. Ogawa Plaza in Oakland, California. The plaza is a 160,000-square-foot public space and is primarily composed of The Commons, a raised lawn. At the helm of Pro Arts, one of the oldest independent art institutions in the Bay Area, I have curated and transgressed the plaza by commissioning interventions, temporary public works, and performance series that catalyze new connections between artist, institution, and the public. I have always been fascinated with this heterotopic place of life: a bustling city and business center, historical site for civic unrest, and home to unhoused communities. FIELD grew out of a palpable excitement Mrin and I both held for life at the plaza. How do we propose new meanings and collective visions for this public place without entangling with the very power that controls it?

Natalia Ivanova Mount, Director,
Pro Arts Gallery and COMMONS

1926   1930   1954   1960

Supermrin   *Collected images from the Condition Reports, Oakland Cultural Heritage Survey, City Planning Department* (images courtesy Oakland City Hall Archives)

The plaza is empty. By that I mean that people are everywhere, yet no one steps into the sprawling field. We are all peripheral. Ants trailing along the edges of a large glass dish.

I break ranks, enter the manicured lawn, and sit by the oak tree that centers the space. A man approaches me. We converse. Over the course of the next few weeks, we become friendly. He has pain, he tells me he is lost, that he has miraculously avoided prison. He has friends at the plaza. We all sit peripherally, facing City Hall, a Beaux-Arts style imperial building—the first civic skyscraper. Two large sculpted American eagles gaze down upon us from its upper-vined fortress, concealing the old county jail. I think about the isolated oak, over one hundred years old. All the trees around it were eliminated when the plaza was redesigned in the early 1990s to enhance the formal character of the space. Later, the tree was fenced in to prevent the public from using its sunken base as a toilet.[1] This move has been largely unsuccessful, and the result is not very appealing. Nonetheless, sitting together, we ignore, perhaps overcome, some of the discomforts of the place.

The simplicity of our conversations over the next few weeks makes me reconsider my approach to public art. In Oakland, like in many other cities in the United States, recordkeeping, mapping, and surveying championed the frenzied search for gold and occupation of lands. The insidious relationship between observing, recording, and controlling sweeps through this history of colonization and staged erasures. These relationships have been seamlessly appointed within the Big-Data-led global present. Twenty-five cameras along the plaza track and compute our bodies. City officials survey us within a repeating, archetypal, and panoptic architecture that has already defined our neural pathways. Oppressive and dry bureaucratic documents written in analytical research styles discuss and redirect the movement of our bodies: who we are, what we will be.

How might this colonial landscape respond to its imperial architectures? When approaching the heavy machinery of state and industry, what are tools of resistance? Could friendship, unmediated presence, undocumented discourse, subjective acts of speaking, and sharing offer reasonable ways of navigating the formidable language of technological precision employed by the state?

Mrin

Mrin invited me into FIELD a year before the pandemic
hit. In our first iteration for the city, we simply proposed that
it pause mowing of the public lawn at the plaza for one season.

It was a simple but controversial proposal. We likened the
routine mowing of the lawn to the repeated and unending acts
of erasure that are systematically executed within public land
in the United States: of histories, of identities, of difference. The
monoculture grasses became our homogenized bodies. We were
interested in unveiling accumulations of trauma and experience
through performances developed in response to the wilding
of the landscape. We speculated an emergent—albeit timid—
diversity within this soil.

Cities invest billions of dollars a year to maintain public
lawns. In the US, it is estimated that lawns comprise more than
three times the acreage of agricultural corn production, making
them the single largest irrigated crop.[2] Large industries
mechanize our domination of grass: mowers, trimmers, seeders,
whackers, irrigators, pesticides, miracle growers, GMO seeds
that never flower. Arrays of sprays and colorants enhance hue,
videos by machine-welding suburban men offer tips on how
to compete with your neighbor's lawn. We are invested in this
state-led aesthetic of control. In a way FIELD was a premonition
of what was to come in the world. The virus turned cities silent
as we were forced indoors. The plaza lay unmowed for the
duration of 2020. Wildflowers began to bloom.

Jess

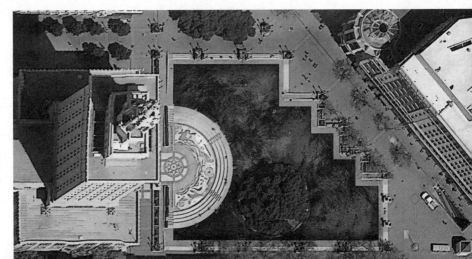

Supermrin
*Frank Ogawa Plaza digital collage
using Google Maps images*

I was surprised when the Public Arts Council of the City of Oakland approved our proposal in January 2020. "What will happen?" they asked. "Will folks think the city is broke? Will they understand?" We outlined the performances that would accompany the land intervention, like a ceremonial lawn mowing at the end of the season. But I was distressed. I had come to love those growing grasses. How might I commemorate them in death?

I met Jil through a connection at Genspace, a nonprofit organization that connects artists with scientists. We spent every weekend for a year working together, masked, two strangers cooking in the studio, developing recipes for biomaterials from waste lawn clippings sourced from the Green-Wood Cemetery in New York. In August 2021, we presented our extensive research at the Climate Provocations exhibition on Governors Island.

Jessica Fertonani Cooke
Braiding Field performance on Governors Island, New York

The material is amazing, versatile. She can be translucent, opaque, structural, skin-like. She smells of compost. She is both dead and living. She likes my attention, likes to be touched. I identify with her invasive, rhizomatic growth patterns, with her fragility. Like her, I am an immigrant—alienated from land, searching for resources amidst precarious circumstances, resilient, able to grow wherever they let me. We make drawings, sculptures, and performances together. She is an active participant and can be demanding and unruly. She doesn't always hand me the reins.

Through a series of public workshops, FIELD grew into a network of womxn that would cook collectively, create, share. Extracting cellulose from waste lawn clippings, fabricating with robots and handmade craft techniques, we have developed a non-Western vernacular and a plant-centric view of the world. Clippings collected from sites across the US are boiled, washed, cooked, layered over river rocks as molds and tree branches for support. These are exhibited indoors as drawing and sculptures, and outdoors as temporary installations in public parks and plazas. Outside, I am interested in the way they decay over time, with exposure to rain, sun, and pollution. Impregnated with native wildflower seeds, these installations introduce new birth cycles in the plaza, generating unexpected encounters between people, birds, park benches, animals, trash cans, weeds, flowers, and buildings.

FIELD artworks transform over time—sagging, aging, drying, seeding, shrinking—representing the vulnerability and fragility of our ecosphere and bodies. While our fears about the future remain real and constant, FIELD has helped me in difficult times, through the communities it has connected me with and the actions it imbues into my body, forcing me to let go of my own desires for control.

Mrin

Supermrin and Xenia Adjoubei   *Material sample using biomaterial,
grass clippings, cow ribs, and performance remains*

The matrescence is a period during which a woman gives birth
to her first child and undergoes the process of becoming a
mother. Although each new mother has a unique bio-chemistry,
triggers, and experiences, the matrescence is widely described
as a tumultuous phase, mixed with the grief of shedding a past
self and the fear of an unknown future.

The slow rewilding of the public lawn must be akin to this
process. Giving birth to and mothering herself may not come
naturally. I make space to listen. I integrate stolen or eradicated
activities into a mixed-blooded present and future. This is
the internal process of decolonization of my body, of her body.
I speculatively develop this matrescence mythology through
collective performances in the plaza. Over the course of a season,
the cosmology—based on acts like sleeping, dreaming, breast-
feeding, singing, digging, mending, tearing, and weaving—
engages the cycles of the lawn in a call to tenderness, mutualism,
and expanded ecology within public life.

Jess

Recipe

We are excited to share FIELD's proprietary recipe with you. The recipes are free and published using a Creative Commons BY-NC-SA license. This means that you can share and adapt our recipes with appropriate attribution and for non-commercial purposes only. While we hope you can learn as much as we have from developing new materials in our kitchens, we ask that you respect our work and refrain from copying it in part or whole without proper attribution or consent. If you would like to partner in the distribution or adaptation of our recipes, we would love to work together.

Developed by Jil Berenblum and Supermrin, 2020

A. From Grass to Pulp

The first recipe focuses on creating grass pulp and extracting cellulose from the grass. We recommend not to use the utensils used for this recipe for normal household cooking.

Materials

Large waste bags
8 quart pan
Running water
Soda ash
Blender
Nylon straining bag/fine mesh food strainer
pH strips to monitor pH levels

STEP 1: Collect grass from your selected site. You can contact the representatives of your nearest park and ask them for their mowing schedule. Usually, the grass clippings are waste materials and park officials are usually happy to share the grass clippings with you.

STEP 2: Dry the grass clippings. Spread them thin and even over a flat and dry surface. The key is to avoid the grass getting moldy so be mindful of keeping the grass in a sunlit and ventilated area. You can skip this step if you are boiling the grass immediately. Drying is only needed to store the grass clippings for a longer period of time.

STEP 3: Boil the grass clippings with soda ash (3 tbsps for one 8 quart pot) for 2 hours. The boiling of grass creates a strong smell and we encourage you to ventilate the kitchen well during the process.

STEP 4: Wash the grass in running water using the nylon straining bag to retain all that good cellulose. Checking pH to ensure the grass reaches neutral levels. The pH scale ranges from 0 to 14, with 7 being neutral.

STEP 5: Using a blender, blend the grass. Add clean water to it to make sure your blender doesn't overheat. It should look like a green smoothie. You can experiment with different blends for different kinds of results.

STEP 6: Strain the blended grass again to remove excess water. You are now left with a beautiful pulpy material that is mostly cellulose. This is the raw material for all the recipes you will work with.

STEP 7: Ensure you refrigerate your grass pulp until use. Biomaterials are only bio because they, like all living things, decay over time. So just like the other vegetables in your fridge, use them within five days or they will spoil. If you want to extend shelf life, you can freeze them, too.

B. From Pulp to Bioplastic

The second recipe focuses on creating grass-based biomaterial from your grass pulp. We recommend not to use the utensils used for this recipe for normal household cooking.

Materials

150g	Grass pulp (Pre-made from grass clippings)
95ml	Carboxymethylcellulose (CMC) 3% solution (a thickening agent used to make jellies)
190ml	Water
50ml	Glycerol
35g	Potato starch
10ml	Vinegar
5ml	Sugar

STEP 1: Weigh out carboxymethyl cellulose (CMC) and water, ratio: 1:3. Mix ingredients together until homogenous. The CMC will gel with time. Refrigerate until use.

STEP 2: Measure out water, CMC solution, and glycerol. Mix together. Add grass pulp to the mixture. Mix well.

STEP 3: Measure out vinegar, potato starch, and sugar. Add them into the pot. Blend all the ingredients once more to ensure a finer finish.

STEP 4: Cook your bioplastic on a stove. Take safety precautions as you would in a kitchen. Stir, ensuring your material cooks evenly and the bottom of the pot doesn't burn. Keep a slow, steady heat. You will start to see the material thicken and curdle. When much of the water has evaporated, and the mixture is sticky and pasty to the touch, you will know that your bioplastic is ready for use.

Keep refrigerated until use. For best results, use immediately. The material will start to thicken within a few days. You can experiment with natural dyes, pigments, molds, and extruders to work with the material. ○

Notes

1. Laleh Behbehanian, The Pre-emption of Resistance: Occupy Oakland and the Evolution of State Power (PhD. diss.), University of California, Berkeley, 2016.
2. C. Milesi, C. D. Elvidge, J. B. Dietz, B. T. Tuttle, R. R. Nemani, and S. W. Running, "A Strategy for Mapping and Modeling the Ecological Effects of US Lawns."

Supermrin *FIELD (leaves) made of biomaterial and turmeric*

Weeds

"The hardest ground we will ever till is that between our ears."

– VAIL DIXON

Every spring, while busy doing other things...

...there was always one weed that would get away from us.

!?!!

Its deep taproot makes it hard to pull out...

Rapistrum rugosum

aka.
Annual bastard cabbage, giant mustard, turnip weed

Origin: Europe, Southern Russia, North Africa, Iraq, Iran, Pakistan

It's super competitive, crowding out other plants and leaving bare soil when it dies back...

BAAAA?

It's edible for stock but as it grows it becomes a thicket that sheep avoid or get lost in...

BAA?

If not slashed, then it leaves a stand of dry, fine fuel up to 2 metres high...

Copious poisoning has bred pesticide resistance, so when we can, we slash, but there's always move to do...

Something Charlie Arnott, regenerative farmer, says makes us think there must be a better way...

I needed to change the paddock between my ears so I could actually do things differently on the ground. I needed to change my attitude and to do that I was needing to ask myself better questions.

JOEL CATCHLOVE 2022

So, we try and ask better questions: What is this plant doing? What function does it serve? Along the way, we discover Vail Dixon, another farmer who has thought a lot about weeds (look for her on YouTube)...

Weeds are a stage of succession between bare and compacted ground and healthy pastures... They heal the land. We often create conditions for weed germination through our management.

What you resist, persists! ... Focus on what you want to create. Focusing on the problem simplifies the system — embrace complexity!

So... ultimately we need to create the conditions for the plants we want to thrive, but in the meantime...

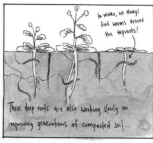
In winter, we always find worms around the taproots!
These deep roots are also working slowly on improving generations of compacted soil...

They flower heavily in spring providing pollen for honey bees...

The thickets they form actually protect our forestry plantings from wind and grazing roos... (Nurse weeds?!)

One of our main projects has been creating habitat for woodland birds...
Seep! Seep! Seep!

...but our first sighting of fairy-wrens came not through our lovingly planted habitat corridors, but through the the wrens commuting from the river to our place via a neighbour's unslashed paddock of giant mustard!
SALTBUSH HEDGE
HABITAT CORRIDOR
WOODLOT

So, it turns out that Rapistrum rugosum is "highly palatable" for goat grazing* - which fits well with our farm plans...
*Search "Weed control using goats"

...as well as being edible for humans! We'll never buy spinach again!

In fact, forager Diego Bonetto (@thecheckyone) made a lacto-ferment from the stuff!
KRAUT WILD MUSTARD

The plants that don't get eaten we can slash to make a reasonable mulch for bare soil over summer...
...and maybe anything we miss mulching we could convert to biochar and return to the soil.

Rather than fixating on the weed, we're trying to appreciate its function in the ecosystem and to understand how it can help us bump the land in the directions we want.
Rapistrum rugosum
← GRASSLAND
WOODLAND →
JOEL CATCHLOVE 2022

Chad Westbrook Hinds

UNTESTED TERROIR

Alpine Winemaking as Climate Adaptation

Locals said we were crazy when we uprooted our small urban winery from the vibrant natural wine scene of the San Francisco Bay Area and set up shop in the decidedly rural western Siskiyou County. The prospect of being able to plant our own vineyards had set a spark that carried us all the way to the northern reaches of the Golden State, but the old timers told us: "You can't grow wine grapes up here! . . . Well, maybe Riesling."

I assured the well-meaning naysayers that Riesling is decidedly a wine grape, and that there are all sorts of exciting and esoteric grape varieties that thrive in the alpine reaches of the Old World. They weren't impressed. The locals remember a time not that long ago when *vitis vinifera* (European wine grapes) probably couldn't have been grown in this small alpine valley. Winters were far more extreme, and late spring snows and frosts were more unpredictable. Today, the only reason we can plant these vineyards is because of climate change.

Because I love the wines of the French and Italian Alps, I spent my winemaking career up until recently trying to channel the alchemy of those distant places' earthy and botanical complexity, but from the Mediterranean climate of Central California. Chatting with a fellow winemaker about my dream of planting the obscure grape Gringet, made somewhat famous by the late Dominique Belluard of France's Savoie region, he laughed and said, "Where the hell are you going to plant that in California? On Mount Shasta?" I couldn't have known it then, but in that moment a seed was planted that steered the direction of our winery—then an urban exploration of natural winemaking—to become a pioneering project in an untested terroir, exploring the potential of oft-overlooked varietals to thrive in the oft-overlooked Alps of California.

Wild grapes thrive in our high elevation alpine valley, in the liminal spaces where the more hospitable valley and untamed forest meet—a lot like witches. The vineyards we are currently planting are being cultivated, but our aim is to mimic nature. We named our first vineyard—an acre and a half of the esoteric light red grape Trousseau—Sylvan, which roughly translates to "from the woods." As opposed to the conventional approach to farming a vineyard, where chemical sprays are used to control the environment and eradicate any unwanted inhabitants, our approach is to enhance diversity through probiotic spray solutions and by planting cover crops that can be found growing alongside wild grapes in the forest. In nature, plants communicate through

Chad Westbrook Hinds *Trousseau vineyard in western Siskiyou County*

a vast underground network, using mycorrhizal fungi like a string in an old-fashioned tin can telephone: transmitting messages and exchanging nutrients with a "for the good of the whole" ethos. From day one, our vines are established alongside mycorrhizal fungi to establish as strong a network among all the plants in the vineyard as possible, and further limit the need for applications during the growing season.

It's a strange feeling to farm a crop newly made viable by climate change. In a moment where we feel somewhat insulated by the changes afoot in our valley, it seems even more obvious than ever that it's essential to plan for

more change in the future. It's in the spirit of thinking ahead that we've expanded on our previously established focus of permaculture to include a goal of dry farming as well. Despite a history of drought in California, grapes have been farmed here without the use of irrigation since the 19th century. Although the trend of commercial growers has been to opt for irrigation to increase yields, climate change is causing a resurgence of dry farming.

Until now, we have planted our vines using as lo-fi a method as possible, sticking cuttings directly into the soil, resulting in an often untested quality enhancement from "own-rooted" vines. The risk of own-rooting vines is primarily to do with susceptibility to pests like phylloxera, a louse which decimated European vineyards in the late 19th century. Because of our very remote locale without any nearby vineyards to spread it, we have seized the opportunity—often praised among European Vignerons of long ago—to make wine from the first commercial vineyard in our area in the purest form possible. Although there was a strong intellectual and stylistic pull to make planting our own-rooted vineyards a trademark of our brand, the alarming state of drought in California has changed our minds.

At least a few years of proper irrigation is normally required to establish a fruit-producing grape plant from own-rooted vines. If we pivot to planting fully established dormant and drought-resistant rootstocks, irrigation demands could require as little water as one good establishing application for the lifetime of the vine. Once the drought-resistant rootstock is established, we can then graft over the vine to the unique alpine varieties from our collection. Planting vineyards with established, drought-resistant rootstocks would reduce our footprint on the land, getting us another step away from active farming and closer to our goal of shepherding a biologically diverse and self-regulating ecosystem.

Farming is a labor of love, and eschewing the many shiny new industrialized ways that can be used to get results can sometimes feel like an uphill battle. With patience, though, every bud break or baby cluster in the vineyard farmed without synthetics feels orders of magnitude more gratifying. Eschewing the status quo of conventional farming is not something I see as a mark of artisanal quality or a decision of luxury; rather, it is a luxury to be able to ignore the effects of climate change,

and with a one-year-old daughter inheriting the world I leave behind, seeking a right relation with the land seems like a minimum barrier for entry to anyone seeking to make a living through farming. The older I get the more obvious it seems that we really have little control over anything in our lives, and nothing more so than nature. There's a real freedom and beauty that comes with embracing the natural chaos of the vineyard, surrendering the human desire to control, and watching the environment co create something vibrant, healthy, and alive with you. ○

American Homes and Gardens *Tomato vines trained against a wall* (image courtesy of Smithsonian Libraries) →

Kacey Stewart

CERASIFORME

Once again, I try
to bring the cherry tomato vine
back in line—
with the stake it keeps pulling over
with the stake it keeps growing away from.

I break the branch
I seek to support—
losing seven near-ripe fruit.

I gain nothing
but knowledge.
Learning to dance
with plants
one step at a time.

Briana Waltman *Perre in the Elderberry*

Ben Carver

THE ONLY DAY

Wrapped securely in a quilt, the birds wake me before the cold. The birds seem to be simply everywhere, and once out of bed, so does the cold. The fire went out in the night. I rekindle it from its warm ashes, place a log in the stove, leave the damper open and door ajar for a minute, and it takes off again.

It's dark as I brew a pot of coffee. When the sun rises over the ridge and hits the window, I put on the stocking cap that grandma made, put on my boots, pat the dogs a hello, and spend the next five minutes starting the quad. Frost on the seat melts into the seat of my pants by the time I get the machine running.

I give the lambs a bale of hay, and a hay flake and grain to the two recently weaned calves in the orchard. I move the sprinklers—all have rings of ice around them—pick nectarines to eat with breakfast, and haul firewood for the stove into the house. The dogs eat the food I give them; I pull a package of elk steaks from the freezer to thaw for dinner. Eating oatmeal with the freshly picked fruit, I notice that one of the bulls is on the road up the canyon.

I hop on the quad again, this time with a dog in tow. I rip up the dirt road, open the gate nearest the bull, and herd him back into the pasture. The dog jumps off the quad and the bull jumps over the fence at a point where its wires are low because the triangle posts have collapsed. I close the gate and resecure the posts with rocks to support the fence before going back to the headquarters.

As the sun rises to meet the bluest part of sky, I grab my ax and chop red fir into pieces small enough to fit in the stove. By noon, I am sore.

I'm deciding what to do for lunch when a car I don't recognize rolls down the long gravel county road and up my driveway, stopping in front of my house. A mustachioed man and a small boy, about six or seven years old, get out. They are both clad in ties. I wish that the dogs would've relentlessly barked at them. I stand with my ax as they address me.

"You live here?"

"Yes."

He extends his hand and tells me his name.

I don't offer to shake hands, mine are sweaty in bloodstained leather gloves and torn up by fir slivers. I keep my eyes locked on his. He takes his hand back and places it on the boy's shoulder, looking at him, "And this is my son."

The kid doesn't look up but mutters a meager hello.

"Young Jacob here wanted to come out on this beautiful autumn day and share something with you," he says. "Isn't that right Jacob?"

Still looking at the ground, the kid holds up a pamphlet and rattles off a memorized speech. I take it from him and knowingly hand it to the man. "I'm not interested. You can pass this on to someone else."

"You're not interested in the truth? With the elections right around the corner a lot of people are saying a lot of things and I know that I hate being lied to. Don't you?"

"Well yeah, it's why I live out here."

"This sure is beautiful country," he says, looking around the canyon. "Driving out here it's easy to see the hand of the creator on everything." Then, looking back to me, "Do you believe in a creator?"

"That ain't none of your business, stranger."

"Alright," he says. "Well, we'll let you get back to your work."

I watch as he and young Jacob get in their car and back out the driveway. The dogs are both laying in the yard and don't seem to notice. Watching the car drive out of the canyon, I notice that the bull is out again. As I had done in the morning, I get on the quad with the dog, ride up to the bull, and he jumps back into the pasture through a different spot in the fence this time. I reset the posts to straighten the top wire, then go back to the house to check the fire and finally eat lunch.

After, I leave to mix a gallon of oil and gasoline to fuel my chainsaw. I hitch the small trailer up to the quad and head off along the creek, toward a pile of scrap wood. Watching for nails, I spend the afternoon sawing up scrap boards into kindling for the kitchen stove. The dogs play and hunt near the water. The sun is nearer to the western horizon now, and I need to begin evening chores. I stack the kindling on the trailer, take the dogs back to the house, and stack the kindling by the kitchen door.

I walk up the hill to the sheep pasture with a bag of offal and feed the livers to the sheep dogs, then come back down. I move the sprinklers, dig up some carrots and beets from the garden for dinner, and pick apples in the orchard to make a pie. The bull is out, again. One more time, I go to him, and he jumps over the fence, again. I fix and raise a different stretch of barbed wires, again.

Back at the house, I start a fire in the cook stove to keep the kitchen warm. For dinner, I grill the elk steaks and roast the beets and carrots. I flip on the television while I eat, but the screen crackles and emits sounds like breaking glass until it makes a loud pop and all of the lights in the house go out. I rummage around until I find candles and light them in the kitchen.

Still wanting to make an apple pie, but without electricity, I decide to use the cook stove. In the candlelight, I peel and core the apples, find the vinegar pastry crust recipe in the old family recipe box, and assemble the pie. I add some kindling to the fire until the thermometer tells me it's ready. For the next forty minutes, I keep an eye on the temperature and add more wood fuel, or open the door to let the heat out, until the pie is done. I have some extra dough, so I bake a quiche for tomorrow's breakfast and let the fire go out. It's still early. The moon is just rising, so I put on a jacket and walk up the creek and listen to the owls and the coyotes and the wind rustling the leaves drying in the trees and gaze disinterestedly at the stars and the Milky Way and clouds passing by.

When I come back to the house I place another log on the fire, crawl in bed with my quilt, and fall asleep almost immediately. ○

Joe Brehm

MAY 1, 2018

Trimble Township, Athens County, Ohio

May it be beautiful before me
May it be beautiful behind me
May it be beautiful below me
May it be beautiful above me
May it be beautiful all around me

—N. Scott Momaday, *House Made of Dawn*

A dozen or so fifth grade students and their teacher stared silently at the bright red Scarlet Tanager perched a few feet above the clear shimmering creek. The bird was uncharacteristically still and easily observed—most bird watchers will tell you it's difficult to get a good look at a Scarlet Tanager because they spend most of their time in the treetops, their bright red and rich black colors obscured by the canopy. With all things, however, there are outliers. Fortunately, they bore witness to this one.

We watched the bird in the mid-canopy. To my amazement, the tanager came down even closer, in the midst of an excited bunch of kids, and perched right along a small tributary of Sunday Creek. They stopped to watch. Some approached it to get a better look, using the slow, cautious stalk of experienced hunters, and the tanager hardly rustled a feather. Curt Moore, their enthusiastic and caring teacher, took his phone out to capture the moment and the bird's striking colors. From downstream

another student came tearing around the creek bend and yelled, "GUYS! We found a crawdad!" The bird flew off, the kids moaned their disapproval, and I chuckled. It was a perfect end to a moment that wouldn't last forever, no matter how patient we were.

This encounter with the tanager is emblematic of the region's assets and challenges. Our region is generally land-rich and cash-poor, simultaneously harboring most of the state's forested lands and some of the most financially challenged school districts. We are rich in outdoor learning opportunities—forests, creeks, fields, farms—but have to work hard for the funding necessary to get kids out there.

The students are bright. A few of them may even remember the bird's name. Most will probably remember that they saw an unusual and beautiful red bird, if only for a second. With time the tanager's red and black colors will merge with their other field trip memories: orange fires started from dried moss, the whites and purples of blooming wildflowers they jumped over, browns of the creek stirred up after they jumped in with their friends, greens of emerging spring foliage, and the yellows of two-lined salamanders caught in the stream.

We live in a region juxtaposed with beauty and hardship; the tanager's beauty will not feed hungry children, repair streams impacted by legacy mining, or cure addiction. But it reminds us that Appalachia is resilient. Wayne National Forest, now harboring breeding tanagers and cohosh, was likely overfarmed then logged several times, yet still recovered. This resilience is also within the humans of this region. After all we are, as John Trudell said, "shapes of the earth."[1] Our hope of a sustainable future in Appalachian Ohio includes the health of our ecosystems. If we can help guide some of these students through their academic journey to a career in ecotourism or natural resources management, the tanager's beauty may actually help them put food on their future tables. There are many wonderful organizations and teachers working hard to make sure we are doing just that.

The moment is special to me because the Scarlet Tanager's beauty is representative of the students' rightful natural heritage. They deserve the opportunity to frolic through big patches of trillium and blue cohosh, to turn over flat rocks in a clear headwater stream and find two-lined salamanders guarding eggs. They deserve to have cerulean warblers breeding amongst the big white oaks and grapevines of their land labs, and to have close encounters with stunning tanagers. They deserve to take field trips that introduce them to the beauty and diversity in their backyards, to know how beautiful their homeland is, and to be proud of it.

The tanager reminds us of the importance of being in awe of natural beauty, and of achieving a state of wonder. This feeling is at the core of what makes us human. Author David James Duncan, in a recent interview with Mark Titus, reminds us that "the goal is simply wonder."[2] Everyone has the capacity and birthright to experience this feeling of intense appreciation for the mysteries of life, no matter our background or socioeconomic status. Regardless of the challenges our region faces, we can't go wrong raising a next generation that knows beauty—in themselves and their surroundings. ○

Notes

1. John Trudell, Public Presentation, Northern Michigan University, January 2003.

2. Mark Titus, "#2—David James Duncan —Award Winning Author," February 22, 2021, in Save What You Love, podcast, MP3 audio.

Bradford Torrey *Everyday Birds: Scarlet Tanager (i. Male, ii. Female)*
(image courtesy of The Library of Congress)

Intercropping

Ethnobotany,

pruning
as science fiction,

experimental food systems

Milo Vella   *Ladder and Young Peach*

Sam Bonney

ON DRAGONS IN THE ORCHARD

An apple tree, in my mind, is a dragon. Its bark flakes in silvery scales, and the watersprouts leaping from gnarled spur-bound limbs bristle like reptilian spines. They are embedded in myth—at once wild and deeply, intrinsically human. In Ursula K. Le Guin's novel *The Tombs of Atuan*, the priestess Tenar inquires of the wizard Ged, "What is a dragonlord?," to which he replies:

> One whom the dragons will speak with. That is a dragonlord, or at least that is the center of the matter. It's not a trick of mastering the dragons, as most people think. Dragons have no masters. The question is always the same with a dragon: Will he talk with you or will he eat you? If you can count upon his doing the former, and not doing the latter, why then you're a dragonlord.[1]

In this sense, pruning an apple tree is dragon riding. The wildlings, the seedlings, and the ancients are trees to be coaxed with gentle pressure and cunning conversation. True, the tools of the trade—razor-sharp pole saws, chainsaws, and secateurs—resemble weapons. But I've learned that fighting is futile and one has no power over the shape of a tree, nor its fruit, unless you speak its language.

In the modern conventional orchard, growers have rightfully decided that they are producers of fruit, not wood. And so they have adopted every trick in the book to subdue the nature of the apple tree. The conversation has become one-sided. Trees are grafted onto university-bred dwarfing rootstocks, producing plants whose only imperative is fruit production. Their roots are so weak that they must be supported on a trellis, irrigated, supplied with artificial fertility, and grown in a competition-free moonscape via herbicides, turning the orchard into a surreal controlled environment vineyard. Limbs are bent, laboratory-derived hormones are applied, and the sacred knowledge of plant physiology—the language of the trees—is perverted in order to forge a tree that is not a tree, and is not a dragon. There are many variations on this theme, but the most common form is known as the tall-spindle orchard system.[2]

By far, the most overwhelming expense in an orchard is labor. Crop load must be managed by diligent, timely winter pruning and blossom thinning in order to assure an adequate return bloom and avoid boom-and-bust cycles of fruit production.[3] Summer pruning permits sunlight to color up fruit for proper grocery store appeal. And of course, fruit must be harvested—a time-consuming process that takes delicate fingers and strong backs. It is precisely to reduce the need for labor that modern orchardists have worked so hard to strip trees of their dignity and wild, dragon nature. Three-dimensional organisms are crammed into two-dimensional "fruiting walls" in order to make orchard tasks legible and streamlined. I want a world where orchards are filled with dragons, like the distant western reaches of Le Guin's Earthsea universe. I have known orchards like this, and for as long as there have been orchards, others have been enraptured by this special world. Indeed, the writhing and mystical power of orchard trees was a favorite subject of artist Vincent van Gogh.

I also sympathize with the bind of the contemporary orchardist. Plummeting commodity prices leave little room for art or conversation with wild nature. As a laborer in a conventional orchard moonlighting as a wild apple steward and forager, I've held this conundrum in the back of my mind while tending to my feral friends, and I've begun to see wild tree forms in a different light. They hint at a new way forward, one that both adapts to the habits of apple-trees-as-dragons and accommodates the need for efficient, standardized production processes.

Apples are a mid-succession pioneer species, often hanging at the edges of forests. They love rich duff, but can't compete in the shade of timber-type trees. I'm always finding them leaning and reaching out towards the sun as their more vigorous neighbors grow up around them. This habit, along with the tendency of their limbs and trunk to bend almost horizontally with the weight of fruit, sometimes results in a tree form defined by long cordons nearly parallel to the ground. From these cordons, vigorous vertical shoots burst from dormant buds, eventually maturing into trunks in their own right. The ultimate effect is a multi-trunked tree like a lopsided candelabra. These multiple trunks disperse the vigor of the tree. Whereas a single-trunked apple can reach heights of thirty or forty feet, these "candelabra trees" are far shorter, perhaps ten or fifteen feet. Small, fruitful limbs branch horizontally off of the multiple vertical trunks, which are in turn arranged in a straight line along the primary horizontal trunk. They are, in my estimation, conspicuously reminiscent of tall-spindle orchard systems, except that in this case the vertical trunks do not exist on individual dwarfing rootstocks but are instead part of a larger, strongly rooted, and free-standing tree.

How hard would it be to bring this candelabra tree form into the orchard? It just might offer all of the two-dimensional legibility, lader-free pedestrian scale, and fruit coloring capacity of modern systems, while negating the need for trellising, irrigation, and herbicides by allowing for standard-sized, deeply rooted stocks. There are kinks to work out, even if it is an entirely sensible shape for *Malus domestica* to take on its own accord. How to achieve something like the early yields of dwarfing rootstocks? Would the labor required for trunk training outweigh the effects of the labor-saving architecture found later in the orchard's life? How will different cultivars respond to such a system? All are important questions, and it's very possible that this is a ridiculous and unfeasible idea. But therein lies the main point of this thought experiment.

Agriculture as it is generally practiced in the global north cannot stay as it is, nor can it return to what it might once have been. If we want a world with dragons in the orchard and faeries in the squash patch—where we can also meet the needs of a burgeoning population

Milo Vella *Thank You, Bo!*

amid a climate crisis—we must be playful. Old binaries no longer serve us, and we must look with open eyes at all possibilities, whether they arise from beautiful, untended nature or from the most unappetizing of industrial agriculture outfits. There are lessons in all of it. To think that one could enter into a relationship with a dragon, to think that one could ask for what one wants and get it without going so far as to break the creature's will and wildness. But if we read between the lines, we may find the magic necessary. And that is what the future demands. ○

Notes

1. Ursula K. Le Guin, *The Tombs of Atuan* (New York, NY: Bantam Books, 1971), 85–86.

2. Many high-density planting schemes have been developed for apple orchards, with intriguing names like "Tall Spindle," "Super Spindle," "V-trellis," "Solaxe," and "Bi-axis." Whereas traditional apple orchards tend to be populated with fifty to one hundred trees per acre, high density systems can have anywhere from five hundred to four thousand trees per acre. These high densities are achieved through the use of dwarfing rootstocks and strict pruning regimes.

3. Many apple cultivars have a habit of "biennial bearing," meaning that they bear fruit heavily in alternate years. This habit is caused largely by the movement of hormones and carbohydrates between different parts of the tree, both of which can be manipulated by growers.

MILLER RASPBERRY.

Canes of strong growth with heavy rich foliage. Very prolific and so hardy as to have
endured a temperature of 25 degrees below zero unharmed. Berries large, very
bright in color, of excellent flavor and the firmest of all Raspberries.
It never fails to produce a heavy crop and picks for a long season.

Teddy Macker

RASPBERRY

Ruby thimble, afternoon's navel, ceremonial hat of the cricket shaman . . . I soft unhold you from your stem, soft unhold your clement, passionate, almost-audible body in my hand, my palm soon wide as my daughter runs heartbreaking to clutch you, to clutch the bell of sleeping sparks, the roe of dawn, the little crumbling chalice of July oblivion.

A version of this profile with the same title was initially published in This World *(White Cloud Press, 2015), and was printed by permission of the author.* ○

Ashley Colby

CRAFT PRODUCTION IS THE FUTURE

The social institutions of our modern era are simultaneously in peril. We face a climate emergency. Our globalized food systems are breaking down with a resultant supply chain crisis. Systems that we believe in have proven to be unstable. What will become the next world order is being planted now, and it is only through experimentation will we discover which version of the world to come will succeed. Like ecosystem succession, our society is reorganizing.

Small-scale production—not just food, but cottage industry, too—has driven economies for most of human history, is accessible to most, and has the potential to form new networks.[1] In my research on communities of practice and alternative or parallel social institutions, I've found that participating in craft production and local economies actually makes people happy. My findings suggest that re-embedding ourselves in a human-scale economy will reconnect us to fundamental values: meaningful work and connection to others and nature. I also believe that local or bioregional economies can help mitigate some of the worst risks of climate change, biodiversity loss, soil erosion, and social unrest.

Our food system crisis is part of a holistic failure tied up with the others of our time:

social, political, financial, and environmental. Policymakers at all levels assume that food systems change ought to be enacted solely through top-down legislation. The way they envision change follows the standard model of social movements: citizens get involved in political movements that put pressure on politicians, who then change laws. But food systems change must be done through a measured process of experimentation and iteration.

In 2015, while researching for my book, *Subsistence Agriculture in the US: Reconnecting to Work, Nature and Community*, I interviewed people in and around Chicago, Illinois, who produce at least half of their own food through some combination of gardening, keeping livestock, tending an orchard or vines, hunting, fishing, cider or beer fermentation, food preservation, and bartering. Though informal subsistence food production like this is common around the world, it is often marginalized in highly developed countries.[2]

I wanted to understand what motivates participation in food production on this scale, so I spoke with sixty different families in urban, suburban, and rural settings around Chicago. I found that small-scale production is a result of the slow collapse of our global industrial

Dorothea Lange *El Monte federal subsistence homesteads. Three-room house seventy dollars and seventy cents monthly. Rent to apply on purchase. Four in family. Father, carpenter, earns seventy dollars monthly. California* (image courtesy of the New York Public Library Digital Collections)

economy. When systems break down, the resulting dysfunction prevents people from meeting their basic needs. As a result, they pursue alternatives and, out of those new arrangements, a new social order is born.

Some began to grow food in community gardens as a form of empowerment and food sovereignty. In the food deserts of Chicago's South Side, most options are unhealthy, processed industrial food sold out of corner stores and gas stations. In order to access produce or fresh eggs, people must produce it themselves—so they have. For others, the crisis was social: there was little trust in state and federal institutions, as well as corporations, to provide them access to quality foods that would not make them sick. They

worried about the chemicals, contamination, mislabeling, and inexplicable health issues that arise from our food system. Self-production became a way to take back control.

This approach of small-scale cottage production harkens back to a method of agriculture lost to most in the modern age. Today, the number of farmers in the United States has dropped below two percent of the total population.[3] After several generations moving from close productive ties to the land and into cities and advanced economies, many people are now experiencing a sense of alienation. This human shift from the traditional, organic, and embedded world to the rational, mechanical, and alienated one is what Max Weber describes as "disenchantment."[4]

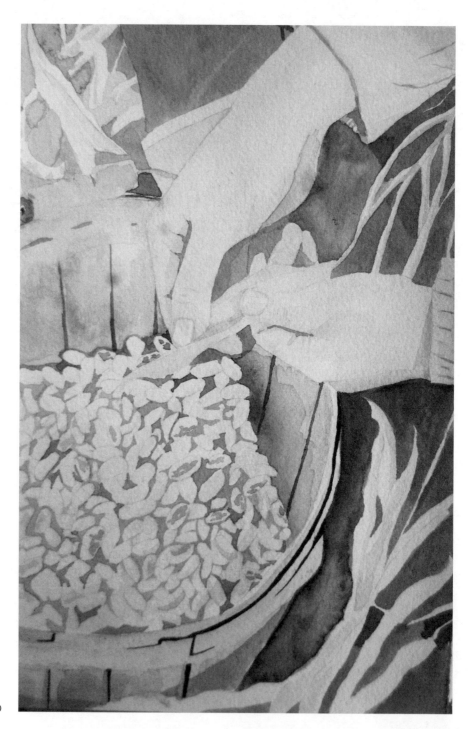

We cannot strive to mirror the past. Instead, we must draw on the best of these eras in tandem with those of modern times to construct a new sustainable social order. While in Chicago, I observed positive change that was iterative, experimental, and innovative, producing worthwhile, systemic results: a connection to nature, meaningful hands-on work, and community.

Some argue that pushing individual solutions to environmental issues is greenwashing, or that it takes the onus off of governments and corporations to address their contributions. What my research suggests is that participating in home food production brings people out of their individual households. Often, in the case of gardening, keeping chickens, or hunting, the act of production requires human troubleshooting and advice. So, in the desire to take part in subsistence food production, relationships are born. I found that individual practitioners become communities of practice.

The strength of weak ties happens when individuals draw on acquaintances for social resources.[5] Groups like the Chicago Chicken Enthusiasts come together to share tips for keeping chickens, barter and trade, re-home flocks, take and teach classes, and host open houses with chicken coop tours. There are communities of practice who have organized politically to protect their newly formed social order, as in one case where a councilwoman attempted to ban keeping chickens in the city.

Ostensibly, these necessary iterative experimentations are the seeds for the next world order being planted. They demonstrate that the process of building entirely new social, political and economic systems is not abrupt: it is gentle, and arises out of continued lessons drawn from the experimentation of solving the problems of collapse.

Frequently, I find a pattern: crisis leads to experimentation which leads to discovery which leads to innovation which leads to creation of something new. There are currently informal communities experimenting in areas including the built environment, architecture, education, transportation, and systems of regeneration. The major barrier to these networks' success is the perception that what they are doing is fringe, or does not fit the mainstream narrative about what a sustainable future looks like. But we face a grand opportunity to usher in the kind of world we want to live in, and the next system cannot be designed top-down. It will only be discovered through our co-creation. ○

Notes

1. ETC Group, *Who Will Feed Us?*, 2017, etcgroup.org/files/files/etc-whowillfee-dus-english-webshare.pdf.
2. Ashley Colby, *Subsistence Agriculture in the US: Reconnecting to Nature, Work and Community* (London: Routledge, 2020).
3. US Department of Agriculture. Economic Research Center. "Ag and Food Sectors and the Economy." 2020. *ers. usda.gov/data-products/ag-and-food-statistics-charting-the-essentials/ag-and-food-sectors-and-the-economy.*
4. Max Weber, *The Sociology of Religion* (Boston: Beacon Press, 1993).
5. Mark S. Grannovetter, "The Strength of Weak Ties," *American Journal of Sociology 78*, no. 6 (May 1973): 1360–1380.

Madeleine Granath

THE HEART CANNOT OUTRUN THIS LAND

Dry place, bleached bone place,
desolation in the ribcage of population ten thousand,
mainstreet plywood like heavy eyelids.
That first year, just after sister's birthday, lightning struck the attic.
We would've lost the house without the garden hose.
Belongings in the grass, sour and black.

Singed cowhide, grandpa's leathered hand coaxes steers through chutes,
the other tying plaits in hemp rope.
We never brand, never castrate before the full moon.
This is before the droughts,
before the stockyard and the auctioneers with their porcelain grins, and
before the pastures are all coyote dens and rattler pits. Before the quail are gone.

The heart is the best meat, this goes for everything.
Hot oil in cast iron, milk and flour,
bite it girl.
This deer heart, shared between grandpa and I,
is when I learn that the heart cannot outrun this land,
never this land.

Omar de Kok-Mercado

A WALK IN THE PARK

I'm standing where the prairie meets the oak savanna pulling out an eighty-year-old fence in thick brush. It is a tedious labor of love to liberate an ancient oak tree whose low-lying limbs enter into the ground and back out again.

Omar de Kok-Mercado *Yellow Coneflower*

The fence and oak tree are entangled with invasive thorny locust, honeysuckle, greenbriar, multiflora rose, buckthorn, and cedar, a result of removing animals and fire from the land. Controlling woody growth in the savanna

Omar de Kok-Mercado
An oak tree in the process of being released

Living in Iowa, I'm no stranger to corn. I grew up on my dad's shoulders in plant-breeding nurseries, beating brown lunch bags full of pollen and transferring them onto eager silk. Breeding for yield, breeding for pest resistance, breeding monotony, and getting a tan. Efficient and productive. In my teens I walked mile-long cornfields pulling tassels to produce seed corn. Soaking wet in the morning from dew and soaking wet in the afternoon from sweat. Diving into the middle of the field to swim in the peaceful quiet, wading in an ocean with no birdsongs, no insects buzzing, no flowers. Looking up at blue sky through rustling corn leaves and staring down at often cracked bare soil. After a long day walking through monotony, I'd drive home imagining the wildness of technicolor prairies and the bustling of wetlands that once covered the glaciated landscape. Summers spent detasseling gave way to time spent imagining, which was my gateway to becoming a soil scientist motivated by a regenerative future.

Today, I have to convince myself that industrial agriculture can be a solution. When the air reeks of Concentrated Animal Feeding Operations (CAFO) manure and brown belters come for the ditches and corn, my faith wavers. When pulling old fence lines on the farm, a nearby plane drops fungicide onto a cornfield, thick like the humid air. Finding gratitude when the mosquitos are thirsty and my nose begins to drip from the chemicals in the air is difficult. I am grateful, and so are the mosquitos.

When thorns and barbed wire tear into my arms and a length of fence is freed from an entanglement of brush, it is this emancipation from an industrial legacy that restores my faith in a land dominated by monocultures. Letting it burn allows hooves to grace the beckoning grass that comes waving back. The next step is

understory mimics the mega-herbivores that once roamed the prairies in what is now known as Iowa. Add fire and this process makes way for fruit- and nut-bearing trees and lush grasslands that support abundant wildlife. Snipping off another length of barbed wire, a squall line approaches the horizon. The whine of a combine harvesting corn is briefly silenced as a thorny locust tree falls into the prairie. These same thorns once deterred the herbivory of Pleistocene mastodons but they don't dissuade a chainsaw-bearing ape in the Anthropocene.

Oak savannas are one of the most diverse and endangered ecosystems in the world and reconstructing them as silvopasture is a path to a climate-resilient future. The industrial production of conventional corn and soybeans dominates the Iowa landscape but the savanna will make a resurgence.

Omar de Kok-Mercado
Monoculture corn dominates the Iowan landscape

building roads of perennials into the row crop matrix to bring livestock back to the land.

Scarlet Tanagers and Indigo Buntings streaking across green oak leaves and flowering prairie plants are like shooting stars. When their colorful birdsongs join the buzzing of pollinators feasting on kaleidoscopic flowers, it can become wildly loud and exhilarating. Bending down into dewy tall grass, the air is

cool and content sheep ruminate along the clear water babbling in the creek at the bottom of the ravine. This nameless creek feeds into the Des Moines River, which feeds into the Mississippi River, which feeds into the Gulf of Mexico. Looking at a hydrology map of the Mississippi River Basin, I see roadways uninhibited by physical barriers, arteries and veins, alveoli, tree branches, fungal hyphae glu-

Omar de Kok-Mercado *A reconstructed tallgrass prairie*

ing soil particles together to form aggregates, the architecture for life. Rivers are completely unlike a fragmented and linear industrial agricultural landscape—and therein lies the solution.

Historically, savannas in the Midwest followed rivers and streams, so there is intrinsic potential in reconstructing the landscape alongside rivers, streams, and roadways as a continuous ultradiverse savanna corridor. As a savanna moves out from the banks of a river and meets cropland, it follows acres not suited for row crop production, which every farm has. Planting and connecting marginal row crop acres with perennials allows for new opportunities and value to people and wildlife. Trees, shrubs, grasses, flowers, nuts, berries, and more to feed us and the land. Thriving perennials connecting marginal cropland may import nutrients to row crop fields by using the savanna corridor as a highway for grazing livestock. A corridor for wildlife is a corridor for people. Regenerative grazing is a public service.

As grazing livestock move across the landscape, mimicking the thirty to sixty million bison that once roamed the Great Plains, they can selectively access cover crops and fertilize fields. These mobile herds bring mobile solar arrays that collect and store water, distribute power back to the grid, provide shade, and serve as Internet of Things hubs for remote sensing technology.[1] Savanna corridors could also connect to existing livestock infrastructure and the manure collected into biogas digesters to produce renewable natural gas.

As a savanna corridor grows into a continuous laneway for livestock, this new form of regenerative traffic will need to be controlled as the animals make their way toward processing and packing facilities. In the digital age, there's no need to invest blood and sweat in putting up and tearing down fences. Virtual fences are now a reality that allows us to open and close virtual gates, set boundaries, and regenerate the landscape without physical barriers.[2]

Look to the world of air transport to validate complex traffic logistics. *Flight to a resilient future is now boarding group A as in Angus.* The total number of passengers carried on scheduled flights now exceeds four billion annually. *Now boarding group B as in Bison.* So, moving one hundred million grazing livestock across North America through savannas could be a walk in the park. *Now boarding group C as in Cooooows.* ○

Notes

1. The Internet of Things (IoT) is a system of interrelated computing devices, mechanical and digital machines, objects, animals or people that are provided with unique identifiers and the ability to transfer data over a network without requiring human-to-human or human-to-device interaction. Examples include connected appliances and wearable health monitors. In the context of the savanna corridor, IoT could verify herd location, monitor plant health, and deploy technologies for livestock and field management.

2. Remote sensing technology refers to the process of detecting and monitoring physical characteristics of the land, like LiDAR, plant biomass, types of plants, checking on livestock herds, finding water, etc. Virtual fence is a tangible technology now as well, it's a solar powered collar that keeps livestock in a virtual boundary.

Rose Robinson

SNAILS GOING TO CHURCH

I imagine attending the snails' mass,
the gathering of soft bodies and shell,
silent hymns calling God towards the grass,
all in their homes, in church, and pools of gel.

There's small reflections of sky in their slime,
the Earth, the sun, and the passing of time.

Frederick P. Nodder   *The Garden Snail*, Helix hortensis

Olivia Brown

AN INTERVIEW WITH GINA RAE LA CERVA

Gina Rae La Cerva is a geographer and environmental anthropologist. Her first book, *Feasting Wild: In Search of the Last Untamed Food*, delves into the relationship between humans and wild foods at a moment when desire for them seems to be at its highest. As the hunger to obtain these foods rises so does their commodification, hashtagging, and over-foraging. What do we lose in the process?

Many wild foods have become inaccessible to their original caretakers, whether endangered from unprecedented demand or driven up in price to unaffordable levels. These foods are left commodified and displaced from the traditional knowledge and culture that surrounded them.

What would it look like to instead cultivate an authentic relationship with wild foods in Western society? What can we learn from the Indigenous peoples that have inherited and passed on their connection to these wild foods for centuries? How can we shift our perspective of humans as separate from the natural world into a perspective of mutuality and reciprocity? What should our relationship to wild foods actually look like?

Gina has foraged wild onions in a Danish cemetery, tracked illegal game meat in the Democratic Republic of Congo, and sipped bird's nest soup in Borneo. From her home in northern New Mexico, she reflected on the lessons and challenges her experiences offered as she traveled the world in search of wild foods. This interview has been edited for clarity.

Why are you interested in researching wild foods?

I've been interested in food and the outdoors for as long as I can remember. I grew up in northern New Mexico, my parents had a garden, and we would forage together and explore in the mountains quite often. As I got older, I turned my interests into academic studies around human-environmental systems and relationships. One day I went to a lecture about cordyceps mushrooms, which grow in the Himalayas. Every season there are tons of people that go up into the mountains to search for them—they're worth a huge amount of money, to the point where the black market is involved. I left the lecture questioning why exactly wild foods have become so expensive and sought-after. At the same time, all my classes were discussing the Anthropocene and the "end of wildness." It seemed like an interesting overlap, that these two things would be happening at the same time. I

Gina Rae La Cerva *Threshing Rice in Borneo*

kept asking myself, "What is this cultural moment really about?" I began to question how we even understand what "wild" means, and to explore our desire for it. I wanted to deconstruct the white colonial view of nature as separate from humans. I like thinking in a complex systems way—in my mind I needed to find out how all of these things connected.

Do you imagine it's possible for the United States to recenter wild foods in a truly authentic way?

What I've come to believe, post writing this book is: wild food is everybody's heritage, we all deserve access to it. But it's tied into so many larger issues around accessibility and privilege. Ultimately, before we can recenter

wild foods in an authentic way, we'll need to ask ourselves how we can cultivate a more values-based food system—a food system that prioritizes more than just productivity. We can learn a lot from how Indigenous peoples cultivated the land pre-colonization. The idea of all of us being able to return to going out to our backyards and collecting our food is unrealistic—but my hope is that this newfound interest in wild foods is inspiring people to say, "Oh, what are the weeds in my backyard? And, oh, 90 percent of these are edible?" We can all connect on that level, we don't need thousands of acres of forest to start eating wild food.

On the other hand, we can't treat wild foods like an Instagram trend. Over-foraging can seriously disrupt the ecosystem. If there's an area you want to forage, walk it for an entire year and don't pick anything. Just notice all of the plants and animals. Notice how it changes over the seasons, what's in abundance and what isn't. Learn about the Indigenous peoples native to that area. Start to build a relationship with the land before you take from it.

Indigenous women's knowledge, especially, is a focus of this work. Why is that?

I cover how Indigenous women's knowledge of wild plants has been plundered by white men in the field of botany. I point out how traditional knowledge was gathered and disconnected from its source—the land and people—then renamed and redescribed through a reductionist viewpoint. Western scientists fragmented the information and stripped it of its complexity.

Are there any lessons that this experience brought you that you could share?

I came to recognize that how we connect with people and places is less about control and more about being in relation to. For instance, how can we think of our relationship to food and the environment as one of mutuality?

Adrien Segal *Wheat Mandala Series:* Puccinia striiformis

What does it mean to be alive on a dying planet? What does it mean to know that every time I eat it is an ecological act, or potentially a violent act, even if I'm doing my best to eat in a sustainable way? Nothing is black and white; everything is on a spectrum and in relation. ○

Gina Rae La Cerva *Fish and Rice*

Sonomi Obinata

ABOUT A DANDELION

In the middle of the cold winter, we farmers get to sit and think thoughts, and imagine the wonderful things we will create when warmth returns with spring. In January 2016, I attended the Pfeiffer Center Midwinter Agricultural Intensive. We focused on the study of two biodynamic preparation plants: yarrow and dandelion. This study inspired me to start what I called the Dandelion Festival, an event on (then) KK Haspel's The Farm.

In our studies, longtime biodynamic practitioner and teacher, Harald Hoven, instructed us to stand next to the dandelion and to observe the plant, first looking at it out of focus then in focus. I asked inwardly, "May I know who you are?"

Reaching out to this plant, I wished nothing but to understand it. I tried to feel it, to perceive what it would tell me. Bright warmth came in, and the plant seemed to say, "I am here to help you—you need us."

I wasn't sure if I was hearing myself speak or if something else was indeed speaking to me. It was one of those moments that we biodynamic farmers experience, we who spend time with our plants doing our best for their growth and always wondering what more can be done. Sometimes something comes into our imagination. I placed those important words aside and continued to study the physical parts of the plant.

Dandelion leaves are solid green with a zigzag outline. Each side of a leaf may be asymmetrical or symmetrical—they do not seem to care for a perfect shape, but rather for the good balance that nature always creates. Each single leaf grows in a spiral, one above the next, taking care that the maximum amount of sunlight is absorbed. They lift themselves up, moving towards the sun. The leaf vein carries within it a beautiful purple tone, and the whole of the leaf is tender and smooth, but also vibrant and solid in texture. It has almost no smell, one that resembles the presence of water.

When I put it in my mouth, it was tasty—not sweet, but bitter.

The flower is very sensitive to the sun. If the day is dark and gloomy, the stem of the flower will not stand. But when it is sunny and warm it stands upright—a long single stem with a bright yellow flower made of hundreds of petals in the center. One can tell that the flower is fresh and full of vigor when there is a bull's-eye in the center. Surrounding the eye are stamens waiting to unfurl and release that sweet, yellow pollen; this dense powder, when touched, will turn your finger yellow. The pollen of the dandelion is important food for

Sonomi Obinata *Sonomi with dandelion painting*

the honeybee after a long, cold winter. It is the first real and plentiful food that they can survive on. In the spring, if you don't yet see the flower, I encourage you to approach the plant and take a peek at the cluster of leaves. Within it you will likely find many potential flowers that look like buttons sitting there ready to burst out and up towards the sky.

The root—oh boy. If you ever try to eradicate the dandelion, you may find it difficult

to do because the taproots go down so deep. It is likely you'll end up leaving some of it in the ground. These long, expressive taproots help to improve the condition of the soil by breaking up compact ground. They are also edible—like any part of the plant—and in fact, can be used as an amazing alternative to coffee. It is best to harvest the roots in the early spring or fall. They must then be rinsed well (a brush may be required in order to remove the soil, though a little soil never hurt anyone), chopped, and roasted in the oven until lightly toasted. Once roasted, the roots must be ground in the same manner as regular coffee—but note that dandelion roots do not contain caffeine. It carries an earthy flavor that gives warmth and tastes bitter like any good coffee does.

Studying the plant at that moment, I was fully aware of this familiar plant whose presence I have seen everywhere in the world. I wondered, "Why are you here on this Earth?"

Where I live now, in Long Island, New York, they start to bloom after the spring equinox and through Easter and Mother's Day. They pop up everywhere—the road side, along vineyard rows, lawns, parking lots, school playgrounds, little cracks between steps. Once I saw one growing out from the bark of a tree trunk.

Elizabeth Blackwell *Dandelion*
(image courtesy of The New York Public Library Digital Collections)

Then a clear vision from the past came in: she was growing out of the crack of hard, flat, gray concrete, no soil around, a blooming, bright yellow flower. I saw her on the way to school as a child.

What does the dandelion mean to you? Weed? Food? Wish-blower? I was struck by what I perceived. When one learns about the dandelion from Deb Soule and Jean David Derreumaux like I did at the Agricultural

Intensive, it is from a medicinal point of view. Intake of the leaf in the spring helps cleanse and support the functioning of the liver and gallbladder as the body comes out of its long winter hibernation. The leaves contain "iron, phosphorus, sodium, and particularly large amounts of calcium and potassium."[1] Interestingly, a recent study showed that dandelions can block spike proteins from binding to the ACE2 cell surface receptors in human lung and kidney cells.[2]

On that day, Mac Mead, director at the Pfeiffer Center, read from the *Agriculture Course* by Rudolf Steiner:

> The innocent, yellow dandelion is a tremendous asset because it mediates between the fine homeopathic distribution of silicic acid in the cosmos, and the silicic acid that is actually used over the whole region. The dandelion is really a kind of messenger from heaven.[3]

Mac told us that dandelions can communicate with each other miles and miles away within a region or on opposite sides of a lake, sending signals to help one another. Plants do not have legs to move around, but there are supersensible forces working here as everywhere, which are not always detectable by modern science.

When I heard, "I am here to help you— you need us," I felt that I was being given wise wisdom whispered from the cosmos.

So, I stood up to answer: "I am listening, and I support you. You continue to return more and more, it seems, even as we call you a weed and attempt to eliminate you for a perfect green lawn. Your life force is so strong to do what plants do—give and give and give, constantly. That is because you are the cure.

I promote you! I will do what I can to help shift people's mindset, to encourage them to refrain from calling you the enemy of the lawn."

At the closing of the Midwinter Agricultural Intensive, I stood up and thanked the teachers, and I announced to the group my intentions for starting a Dandelion Festival. We've done it every year since, until COVID-19 concerns, with hundreds of visitors annually.

Support has been wide-ranging. Alex Tuchman at Spikenard Honeybee Sanctuary in Floyd, Virginia joined in this movement and started a Dandelion Celebration in 2021. In 2022, Churchtown Dairy Farm in Hudson, New York, started its Dandelion Festival, which it plans to hold annually. Just as I intended, these celebrations are expanding. They are awareness events for when the Earth awakens in the Northern Hemisphere, and encourages a sensitivity to this resurrection. ○

A version of this article was initially published on the Biodynamics Association Blog *under the title* "Let's Start a Dandelion Festival!" *and was printed by permission of the author.*

Notes

1. Doug Elliott, *Wild Roots: A Forager's Guide to the Wild Edible and Medicinal Roots, Tubers, Corms, and Rhizomes of North America* (Rochester: Healing Arts Press, 1995), 71.

2. Hoai Thi Thu Tran, Michael Gigl, Nguyen Phan Khoi Le, Corinna Dawid and Evelyn Lamy, "In Vitro Effect of Taraxacum officinale Leaf Aqueous Extract on the Interaction Between ACE2 Cell Surface Receptor and SARS-CoV-2 Spike Protein D614 and Four Mutants," Pharmaceuticals 14, no. 10 (October 2021).

3. Rudolf Steiner, *Agriculture: Spiritual Foundations for the Renewal of Agriculture*, trans. Catherine E. Creeger (East Troy: Biodynamic Association, 1993), 103–104.

Hans Kern *Escape from Scare City*

Adrien Segal *Wheat Mandala Series: Man-Made Famine*

Can, will do!

Harvest season energetics,

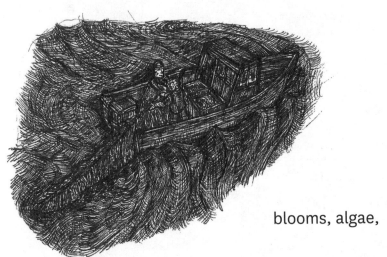

blooms, algae,

and other excesses

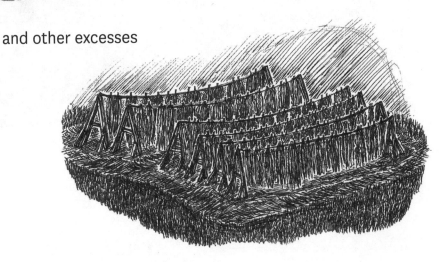

Samantha Winship

THE RHYTHM OF MOTHER AND HER BEES

Rose Robinson   *The Chatterbox—Putnam Camp: Keene Valley, New York*

Imagine North Carolina without sweet potato season. The tuberous roots have a rich history in the area, and were cultivated by the region's original Indigenous peoples. As of 2017, the state produces more than 40 percent of the national supply of sweet potatoes.[1] As of 2021, the state was ranked first in fresh sweet potato shipments.[2]

The Indigenous practices of my family are imprinted into me. I can remember the fond memories of my grandmother growing up in Goldsboro, North Carolina, prepping for Thanksgiving. It was truly an adventure to travel with my grandmother around to pick up sweet potatoes from the farmers she knew for her delicious sweet potato pies. She would send every single one of her nine kids home with a pie for their family to enjoy later. Watching her bake was so special and the smell still lingers in my heart and mind. The love that it took was truly amazing and something that will make me forever cherish how important traditions are.

In my family, I was the first beekeeper my grandmother met. It made me feel so proud. I always felt she had a sense of knowing what we needed and that was truly a gift, along with how she used food and love, and how she shared it with the community—just as I am doing today.

As a farmer, I see the effects of climate change happening around me daily. Climate change is drastically changing how farmers experience each growing season. I became a farmer and beekeeper a few years before 2019, when the COVID-19 pandemic hit. Tuning in to the land has been eye-opening. I did not realize how important my journey seeking self-suffi-ciency for me and my family would be. There are so many lessons in farming and beekeeping. As a steward to the land I see that what we do today has a direct impact on the future of the world around us. The work and the life cycle of the bee is a small act that proves this be true.

The life cycle of the worker bee is roughly around eighteen to twenty-two days. It is estimated during that life span that they may visit up to one hundred flowers during a foraging journey and make up to ten to fifteen trips per day.[3] It is simply amazing, the work bees do. There is a little magic inside beekeep-ing. That is an idea that I like to share with my son Kingston when we are out in our bee yard. Learning all the magical things bees do has been an important part of his homeschool classroom, which my husband and I decided to implement a year before the pandemic. My bee yard became a part of my sanctuary and a place of refuge for me and my family

Samantha Winship
Preparing for beekeeping

during the pandemic. Just like everyone else I was trying to find a sense of normalcy in an ever-changing world.

I first introduced Kingston to the bee yard when he was five. I remember taking him to go pick out this tiny little bee suit and his eyes lit up when we drove home with our first package of bees. I enjoyed watching him learn and explore, and it was amazing to see each and every time just how brave he was to work alongside me and the bees, from getting his first sting to holding his first frame on his own. Now, he's even helping me catch swarms. Our connection to nature and the value of being with each other is the center of what we needed to get through the pandemic. Interconnection—to humans, nature, and that magic of the bees—wrapped its arms around us. Through a period of uncertainty I found this new passion and now I have realized they are also keeping me, in more ways than one.

My bee yard and all the elements around it are my sanctuary, built during what felt like a fast-moving storm. The global pandemic forever changed how I view the world around me. I still embrace it and share as much of the beauty I found along the way with the rest of the world. I am empowered that the energy of the bees surrounds me, the flapping wings of the monarch butterflies guide me, and the sunshine shines down on me and forges the way for a brighter future, with more and more people learning to enjoy the simple things. Get out into nature and remember to pay attention to all of the beautiful things that surround you. There is beauty in the struggle. And it will be stories that are passed onto future generations and this story of triumph and empowerment and learning how to *just bee* will be one of them. ○

Notes

1. Minda Daughtry, "North Carolina Sweet Potatoes," N.C. Cooperative Extension, last modified September 7, 2021.
2. Tom Karst, "North Carolina suppliers continue sweet potato dominance. The Packer, August 12, 2022. *thepacker. com/news/produce-crops/north-carolina-suppliers-continue-sweet-po-tato-dominance.*
3. Charlotte Anderson, "Complete Honey Bee Life Cycle," Bee Farm (blog), *Carolina Honeybees*, August 16, 2022.

Danielle Walczak

AUGUST

Giving In

> **Each body whose**
> **wings become invisible,**
> **Squints toward the horizon.**
>
> —Ali Briere, "Geese to Brick Island"[1]

August is nagging tomato vines tickling our shoulders, waiting to be pruned. The long warm days and short warm nights send the tomatoes ripening faster than we can sell or eat them.

Here, the tomato's magnificence blurs from individual splendor to sheer volume. Forty-five gallons of cherry tomatoes a week, thousands of pounds of heirlooms and red slicing tomatoes. The steady flow in previous months breaks into a flood of fruit.

August is about surrendering. Swimming downstream, letting our bodies do the work. From here, there is only harvesting, and we rely on each hour of sun. My body remembers the rhythm and submits to the movement, not new to the tension of summer skin over muscles, the prick of vegetable-based dyes. Joy is simplified during this time: finding refuge under a shady tree, the splendor of an orb-weaver spider, the comfort of repetition. The comfort of seasonal returns despite a disorienting pandemic.

Machine thinking is the opposite of mindfulness, I repeat in my head.

But I said this was about surrendering.

As is routine every Thursday—"the Tomato Sabbath"—we kneel in dew-covered rows of verdant umbellifers. Parsnips, carrots, celery, bow from their altars into the walkways, offering harvest. We pull their leathery tap roots from the soil, trying our best not to crush their fronds.

I remember my first season farming, dancing like a field of carrot tops in the 2:00 p.m. bay breeze, brutishly and benevolently clawing my way through parsnips, smearing their sap all over my arms. Later, after being in the humid sun, blisters formed under the sap. Phytophotodermatitis is the name, burning is the experience. The scars remind me. My body is not new to this seasonal falling and rising.

The next day, and every week, we crouch and crawl on our "Friday knees" beneath the canopy of ten-foot-tall tomato plants to do the big weekly harvest of tomatoes. It's six in the morning and, at the base of the plants, the cooler night air still lingers. So do the mosquitoes. Warded off by the early summer drought, it seems their delay only doubled their fortitude. They quietly sneak under my mask and sting my upper lip, the small strip of visible skin on my back and ankles, between pants and shirt and shoes painted in dried blood. Still, we keep picking.

The quiet morning fills with profanity and the trays are filled each with fifteen pounds of heirloom tomatoes. The sun eventually finds us here. Sweat runs in my eyes and I decide the salinity is more comfortable than a tomato-tarred finger to the eye. The six of us harvest for six more hours. We laugh at how ridiculous it all is, like pigs covered in mud, rich in vegetables—feasting.

Every Friday, Ali and I camp next to the field of wildflowers by the bay. We sleep in the small buffer of sumac and ash along the bank separating the water from the fields. I sit crouched under my blanket to eat a mosquito-free dinner. My back, the ridgepole for my tent. I make noises like Gollum and Ali chuckles; I'm surprised to hear myself join her. These days,

my body can hardly allow the release of laughter. An odd reminder of the spontaneity of life before COVID-19. How not everything, but some things, were easier.

Turning like the tides in the bay, the wild-flowers near our camp spot change almost every week. The surge of lupine, St. John's wort, milkweed, rabbitfoot clover, black-eyed Susan, goldenrod. Now, filled with Queen Anne's lace.

The August air is potent and sweet and cools once we step down off the bank. We sit on a driftwood tree by the water.

"The marsh chill, the smell of low-tide desiccation," Ali notes.

Wild rice-like fans of daggers piercing the saffron sky. Ali squints to the horizon, talks about getting a tattoo of the brimming white cup growing next to her. She examines the fibers that make a flower in her hand before bringing it to her lips, as if to smell, but instead takes a bite.

"Umbellifer," she declares, unsurprised. "I knew it."

Sometimes, the closer you hold that which hurts you, the more in control you feel. However, she tosses the flower back to the outgoing tide. Giving into the evening what she already knows about burns and seasons. The sun dips behind the horizon, and its ability to turn mastication and skin into fire is gone at least for now. Looseness is only in the sleepy delirium before bed. For a moment, we feel safe.

Over a field of Queen Anne's lace reflecting the moon's light, we settle in. We watch the pulse of a firefly: blatant and alarming up close, soft and spectacular amongst its kin. ○

Note

1. Ali Briere, *Poembody, Somebody* (self-published, 2021).

Briana Waltman *Raking Blueberries*

Nina Montenegro *Harvesters*

Emily C-D   Trueque Translocal de Semillas Libres *(Translocal Free Seed Swap)*

Mariee Siou

MAÍZ, LA LUZ DE LA MADRE

Maíz, la luz de la madre
Resting in the power of reaping the holy
Ordinary in the eye of God
Pollen in the star of my mother
Precious love we lay here
Folded in husks at the foot of Earth's alter
Summer of walking the maze
A puzzle of cobs
Children, kernels of rainbows
Green seas part in remembrance
The inheritance of fields
Sunlight ground through millennia
On the stone of great-grandmother's metate
Blue corn sewn by the thread of man
A maiden's ear listening in a bed of soft tassel
Masa kneaded by an empire's hands
Golden tea of silks
Drinking atolé from red clay bowls
Flower antennae to the heavens
Speaks in the tongue of stars
Secrets shared with kin
Wildly shuck—find the sun
Cradled within

Linley Dixon

ORGANIC MOTHER TREES

In the mid-1990's, scientist Suzi Simard discovered that nutrients are transferred between the roots of forest trees through their mycorrhizal fungal connections. Her research, published in the distinguished journal *Nature,* challenged a government policy strategically named "Free to Grow" that required herbicide use on "reforested" clear-cut land.[1] Under this policy, tax dollars were funneled to chemical companies to poison public lands.

The prevailing paradigm was that after logging, herbicides should be sprayed under "reforested" fir trees. It was thought that the newly planted fir saplings would thrive if there was no competition from surrounding "weed" species. But Simard was outspoken in her observations that reforestation was more successful if the native biodiversity in the understory was allowed to stay. As such, she was ridiculed by both the scientific and forestry communities for suggesting that cooperation between species was a greater force than competition in nature. Now, vindicated by many in her field who publish work that supports hers, she is celebrated.

Organic farmers have always observed and intuited what Simard's data demonstrates: surrounding vegetation provides ecosystem services and also can be a friend to neighboring crops. While we don't want weeds to dwarf crops, organic farmers often see them for the benefits they bring: mining nutrients from deep below, enticing pollinators, or confusing pests.

Simard later researched what she calls "mother trees," the elders of the forest that feed the neighboring saplings of many different species when they are young and shaded. Through mycorrhizal connections and root grafts, elder trees provide nutrients to young trees that are struggling to compete.[2] In the organic farming community, there are many such mother trees as well—elders with

Linley Dixon *Mother Trees: Becky Weed with her sheep guardian, Ruby, at Thirteen Mile Farm in Belgrade, Montana*

immense knowledge of organic farming systems, and their local soils, microclimates, and pests. They are the farmers that willingly provide help to the youngers who are just getting started. Like saplings, young farmers have so many obstacles to overcome: access to land, markets, and farming skills among them. If we are to achieve the healthy, local food system we all dream of, we need elders to lend a helping hand.

Farm visits I've taken for the Real Organic Project have shown me that, across the country, the organic movement is still strong and integrity is high. Support and cooperation between farmers is evident. I have met several mother trees, elders that extend a hand to the next generation, and even more hard-working saplings, youngers making a go of it. At the Real Organic Project, we call the areas where many small and mid-scale organic farms thrive "islands of sanity." They are rich in community, knowledge, shared resources, and real organic food that is regionally sourced, nutritious, and tasty.

In these thriving organic landscapes, the failures of the United States Department of Agriculture (USDA) to uphold the true meaning of organic feel somewhat distant. But even in these islands of sanity, there are signs of USDA

Linley Dixon *Mother Trees: Real Organic Project Co-Director Linley Dixon with Jessica McAleese at Swift River Farm in Salmon, Idaho*

failures nonetheless. Farmers describe lower prices, especially in grains, fewer organic dairies, and certified organic hydroponic cucumbers, tomatoes, peppers and greens, imported from great distances, on the grocery store shelves— even when local farmers are at "peak season." As Hugh Kent explains in his Real Organic Podcast interview, the allowance of cheaper practices that don't adhere to organic principles of soil health, actually mandates those practices under the seal.[3] The Real Organic farmers pasturing livestock and fostering soil health simply can't compete with confinement and soilless production. In spite of all the gains we have made as an organic community over the last fifty years, we must be aware that even our islands of sanity are precious and must be protected from the lobbying efforts of corporations trying to profit from the word "organic" without adhering to its principles.

Now more than ever, we need our elders to act like mother trees and extend their reach beyond their local communities to protect their life's work and keep the organic movement strong. When the USDA goes astray from organic principles, it is even more important that farmers in these islands where the local organic landscape appears strong, rise up to protect organic at the national level. After all, we are all connected. As Oklahoma farmer Emily Oakley said about the Real Organic Project in her podcast episode, "Vermont can do this without Oklahoma, but Oklahoma can't do this without Vermont."[4]

We need our so-called mother trees to spread their roots far so that we all can weather the storm. Lest we end up with an "organic" monoculture of livestock in confinement and industrial hydroponic production, much like the isolated "reforested" firs growing alone in poisoned soil. ○

Kristin Leachman   *Bridalveil Fall (Yosemite, California)*

Notes

1. Suzanne W. Simard, David A. Perry, Melanie D. Jones, David D. Myrold, Daniel M. Durall, and Randy Molina, "Net Transfer of Carbon Between Ecto-mycorrhizal Tree Species in the Field," *Nature* 388 (August 1997): 579–582.
2. Suzanne Simar, *Finding the Mother Tree: Discovering the Wisdom of the Forest*, (New York; Knopf).
3. Real Organic Project, "Hugh Kent: Allowing Hydroponic Berries into Organic is a Mandate," July 27, 2021, in *The Real Organic Podcast*, podcast, MP3 audio, 54:02.
4. Real Organic Project, "Emily Oakley: Living Next to a CAFO While Serving on the NOSB," September 3, 2021, in *The Real Organic Podcast*, podcast, MP3 audio, 1:00:18.

Lynnell Edwards

FARM INVENTORY

Ars Poetica

It comes down to this: the mower
won't load and the loader won't mow
and you might need one or the other -
you just never know. And the big
tractor he bought with his dad?
That's sentimental. And the farm truck?
Well you always need a flatbed. And stacked
and stored in the sheds and barns
are sledges and rakes and axes (and bundled
ax handles); an adz and a mattox
bales of wire, buckets of bolts,
five plows and a harrow that once pulled
behind horses, wagons and tractors;
hacksaws and bow saws, a two-man saw;
a vice and grinder and nine pitchforks
leaned against the front wall of the shed:
one of them taller than
a man standing, from days when
making the hay was a dying art,
a lost art, or so said the hired man
as Frost wrote it down. Robert Frost –
a failed farmer and worse teacher

but who might have seen the point
of owning both a mower and a loader,
keeping a tractor for sentimental reasons
just to write a good poem, if not
ever to make the hay or cut the tobacco
or pull a honed plow behind a horse.
So don't tell me what would make
a good poem; don't remind me
to write what I know. I've got
a barn full of sharp tools and not
a one of them metaphorical.

Rimona Eskayo   *Life and Death and In-Between*   235

HI JASON, I WANTED TO ASK IF MY SON COULD WORK FOR YOU ON THE FARM THIS SUMMER.

SURE, I'D BE GLAD TO HAVE SOME HELP.

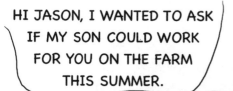

I HAVE SUCH GREAT MEMORIES OF WORKING ON A FARM AS A YOUNG MAN – BUCKING HAY IN THE HOT SUN, SHOVELING OUT THE BARN STALLS, HEADING OUT ON HORSEBACK TO CHECK THE HERD WITH MY BUDDIES...

I'D HATE FOR MY SON TO MISS OUT ON EXPERIENCES LIKE THAT.

SO WHAT ARE YOU WORKING ON THESE DAYS?

WELL, THE FARM REALLY NEEDS A NEW WEBSITE...

Chuck Monax   *New Website*

Chuck Monax *Self-Reliance*

Christopher Winslow

LAKE ERIE BLOOMS

In the late 1960s, Lake Erie—the shallowest, warmest, and southernmost of the Great Lakes—was considered a dead lake. Industry-associated chemical pollution along the shore and within the eight large tributaries with navigable shipping channels were a primary cause, contaminants including carcinogenic and neurotoxic heavy metals, PAHs, PCBs, and other compounds.[1]

This dead lake image was tied to a loss of fishes and aquatic insects. Through an algae lens, however, the lake was too alive. Algal blooms were caused by phosphorus in laundry detergent and the release of phosphorus from old wastewater treatment plants (WWTPs), which were not designed to remove this nutrient. This dead lake was typified by the burning of Cuyahoga River at least thirteen times, most famously on June 22, 1969.[2]

Highlighted in *Time*, this fire elevated the issue, making it arguably the United States's poster child for industrial pollution and the need for federal regulatory intervention. In fact, early editions of the children's book *The Lorax* stated that because of pollution, humming fish were leaving the pond "in search of some water that isn't so smeary." Dr. Seuss rhymed that with, "I hear things are just as bad up in Lake Erie."

The fire in part motivated the formation of the Environmental Protection Agency (EPA) in 1970 and the signing of the Clean Water Act (CWA) two years later. Because of these efforts, Lake Erie soon became an example of ecosystem recovery. Using the CWA as the playbook, the Ohio EPA nearly eliminated discharge of legacy contaminants.[3] Federal legislation banned phosphorus in laundry detergents and required WWTPs to implement phosphorus removal approaches and to upgrade facilities, with federal investment to support these upgrades. Flammable debris and oil slicks in the lake became less common, fish species returned, and algal blooms declined in prevalence and size in the early 1980s and through the mid-1990s. The lake was not instantly unblemished, as legacy contaminants in sediments were still present and habitat sufficient for insects and fishes was lacking, but rampant, unchecked industrial pollution and extreme algal growth seemed under control.

Unfortunately, harmful algal blooms (HABS) within Lake Erie, specifically the Western Basin, returned in the early 2000s and have continued impacting recreational uses and complicating access to clean drinking water.[4] This contemporary algal bloom issue differs in two critical ways from the dead lake blooms of decades prior: it

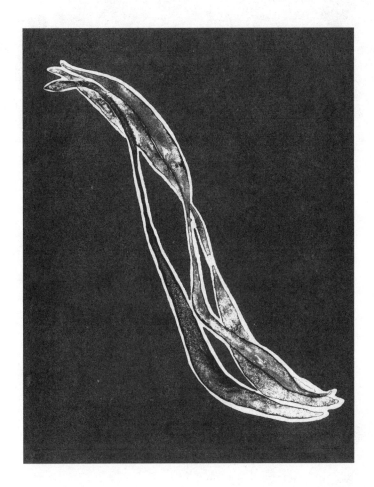

is dominated by a new strain of cyanobacteria—*Microcystis* instead of *Aphanizomenon*—and perhaps most importantly, the nutrient source is different. Today's bloom is primarily driven by nutrient runoff from the Maumee River watershed (non-point source), not WWTPs, the point source of the past. Agriculture production contributes the largest share of nutrients in the Maumee River now (88 to 93 percent), compared to WWTPs (~8 percent), home sewage treatment systems (septic tanks and leach beds; ~4 percent), combined sewer overflows (~1 percent), and urban runoff (lawn fertilizer).[5] Because of this new source—nutrient loss from agricultural lands into watersheds, an upstream watershed contribution—there is a need to educate and engage people that live far away from the impacted lake.

Many efforts have been deployed to address HABs in Lake Erie. Funds have been made available, and agencies, local governments,and academic institutions have played key roles in

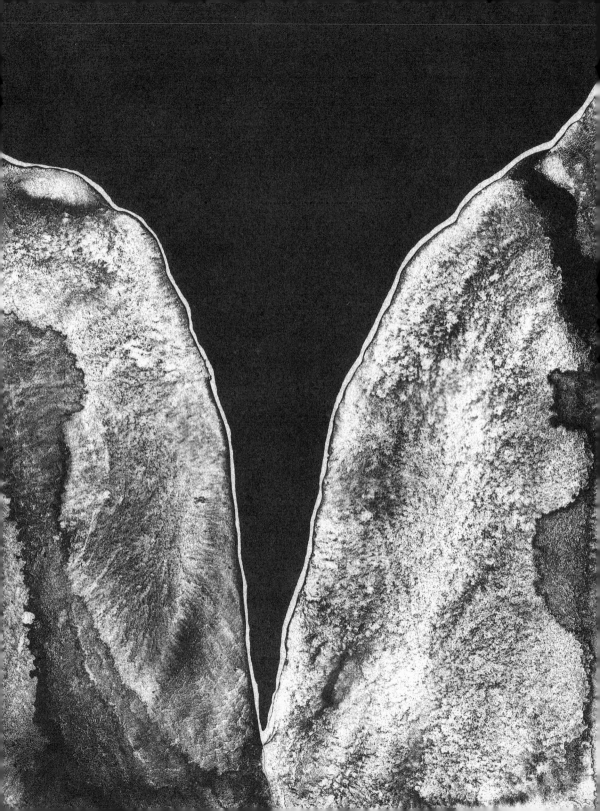

monitoring and implementing mitigation and treatment efforts. Residents of impacted communities, non-governmental organizations, and private industries have engaged in nutrient reduction efforts. Research targets ways to produce safe drinking water and assess possible human health risk and ecological impacts. Researchers are working to understand how blooms behave—when they grow, when they become toxic, and what conditions are conducive to growth—and how to identify ways to reduce nutrient loss from watersheds that drive these blooms. The latter has focused on loss of nutrients from agricultural lands.

Academics and both state and federal agencies are looking to slow nutrient runoff by assessing how to better manage water movement, especially under current and future climate change scenarios. To do so, they work to understand the decisions farmers typically make regarding phosphorus application in a given planting season, including rate, timing of application, type of fertilizer used, and where fertilizer is placed (broadcast application versus banding and injecting), and how to address our previous use of phosphorus and nitrogen fertilizers (often referred to as contemporary versus legacy phosphorus release).

More is known about nutrient application and loss today than was forty years ago. So why is progress toward "recovery"—reduction in phosphorus inputs that drive blooms—slower than efforts to address blooms in the 1960s and 1970s? Communities living on Lake Erie and along rivers with blooms rely on this lake for both recreation and drinking water and want this degraded system to be restored. These communities want the problem resolved as quickly as they feel the burning river issue was resolved. Unfortunately, folks impacted by these Lake Erie harmful blooms do not live in the watershed where these nutrients originate, and do not understand the watershed land use that contributes the bulk of nutrients downstream: farming. They do not understand the decisions farmers are forced to make, the complex economic pressures they face, and farming's complex history. Many do not understand the complexities that impact the rate of recovery.

One complexity is the intentional, although not malicious, excess application of fertilizers in the past. Four decades ago, producers, departments of agriculture and land grant institutions assumed that nutrients applied in a given year stayed in soils for access in later planting seasons. As such, farmers that had high-yield and high-profit years, were encouraged to invest in their soils by applying fertilizer that could be used by crops in future lean years.[6] We now know that high nutrient concentration fields that received fertilizers applications above agronomic rate are often chronic sources of nutrient losses today. These "legacy fields" were not created under malicious intent; producers and academics simply did not fully understand phosphorus movement and phosphorus storage within soils as much as we do today. Addressing nutrient contributions from these spaces, often called "hotspots," will not be easy. They are difficult to locate and identify, and are often financially difficult to remedy, construction of wetlands and/or phosphorus traps placed at the edge of fields to capture chronic nutrient losses.

A second complexity impacting recovery is that agriculture solution sets identified to work in demonstration/pilot study fields or that have been demonstrated to work in academic laboratories are difficult to deploy in the real world. Some of these demonstration farm and lab solutions are cost prohibitive for farmers. For example, toolbars and tractors required to band or inject fertilizers, allowing farmers to avoid broadcast application, are very expensive. Cover crops, which often have no commercial value, cost money to purchase, seed, and burn down. Some solution sets/practices will be new to producers, like cover crops, drainage water management, conservation tillage, subsurface placement of nutrients, and learning new habits are difficult and require farmers to abandon approaches that have been handed down from generation to generation. This learning curve is not insurmountable, but it will take time for land-grant university extension educators and certified crop advisors to train producers and ensure confidence in new approaches.

Some of these solution sets will need to be targeted at specific locations. Land-grant institutions and agencies have come to recognize that no two fields are identical. Monitoring streams and fields that border streams has shown that not all fields are leaching/losing nutrients at the same rate. Nutrient loss can vary because of many factors, including slope, soil type, cropping history, micronutrient differences, and microbial communities in the soil—many of which are not fully understood. This means that some producers are asked to take on a greater nutrient loss prevention burden than others. Finally, there isn't enough monitoring data to know exactly where these hotspot fields are located, meaning farmers will need to self-identify their nutrient loss—and also need to have the desire to engage in activities that slow their nutrient loss.

A third complexity is that approximately half of farmed land within the Maumee River watershed is rented.[7] This has the potential to impede nutrient reduction efforts because the willingness of a farmer leasing land to deploy costly management practices will likely be lower on land that they may not farm in the future.

A fourth complexity that influences recovery is climate change. Climate change trends make solution sets difficult to deploy and complicate the movement of soils and nutrients below the surface that may be mobilized by tile drainage (piping placed under fields to help move excess water off fields). Heavy precipitation that occurs in the fall and spring in northwest Ohio is predicted to increase under most climate change scenarios, overlapping with when nutrient application often occurs.[8] Subsurface nutrient loss is an issue within the upper Maumee River basin as some fields need to be aggressively drained via tiles to grow crops because many are low-lying or were historic wetlands. This means that any rain that falls is moved off fields quickly via necessary drain tiles. Tilling in this region coupled with climate change forecasts means that farmers need to find a way to slow water movement while not impacting yields.

A nutrient target has been identified that will result in recovery: a 40 percent reduction in phosphorus entering Lake Erie via the Maumee River relative to 2008. When we hit this target, blooms considered excessive should only occur one out of ten years.[9] This target can be reached, but it will take time as deploying needed efforts will need to reach more than 70 percent of agricultural acres and many deployed practices will have a lag effect. Further, achieving this reduction will require producers, agencies, land-grant institutions, and elected officials to address legacy, high-nutrient fields that chronically release nutrients, incorporate climate change scenarios into our decisions, and modify our relationship with phosphorus application in a given year. Producers all need to apply nutrients at the right time (avoiding excess rainfall), at the right rate (not applying nutrients when sufficient amounts are already in soils), and in the right place (subsurface, not broadcast application). ○

Notes

1. An Li, Jiehong Guo, Zhuona Li, Tian Lin, Shanshan Zhou, Huan He, Prabha Ranansinghe, Neil C. Sturchio, Karl J. Rockne, and John P. Giesy, "Legacy Polychlorinated Organic Pollutants in the Sediment of the Great Lakes," *Journal of Great Lakes Research* 44 (March 2018): 682–692.

2. Lorraine Boissoneault, "The Cuyahoga River Caught Fire at Least a Dozen Times, but No One Cared Until 1969," *Smithsonian Magazine*, June 19, 2019.

3. Chuanlong Zhou, James Pagano, Bernard A. Crimmins, Philip K. Hopke, Michael S. Milligan, Elizabeth W. Murphy, and Thomas M. Holsen, "Polychlorinated Biphenyls and Organochlorine Pesticides Concentration Patterns and Trends in Top Predator Fish of Laurentian Great Lakes From 1999 to 2014," *Journal of Great Lakes Research* 44, no. 4 (August 2018): 716–724.

4. Robyn S. Wilson, Margaret A. Beetstra, Jeffrey M. Reutter, Gail Hesse, Kristen M. DeVanna Fussell, Laura T. Johnson, Kevin W. King, Gregory A. LaBarge, Jay F. Martin, and Christopher Winslow, "Commentary: Achieving Phosphorus Reduction Targets for Lake Erie," *Journal of Great Lakes Research* 45, (November 2018): 4–11.

5. Wilson.

6. Wilson.

7. *Ohio: State and County Data* (United States Department of Agriculture, 2014), *agcensus.library.cornell.edu/wp-content/uploads/2012-Ohio-ohv1-1.pdf.*

8. Aaron B. Wilson, Alvaro Avila-Diaz, Lais F. Oliveira, Cristian F. Zuluaga, and Bryan Mark, "Climate Extremes and Their Impacts on Agriculture Across the Eastern Corn Belt Region of the US," *Weather and Climate Extremes* 37 (June 2022): 1–22.

9. Wilson.

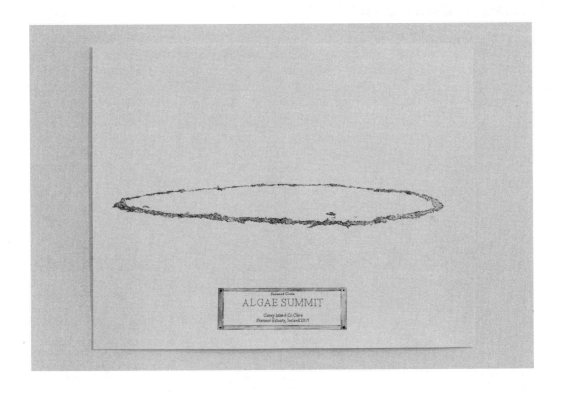

↑↗ **Filip Van Dingenen** *Algae Diplomacy—Almanac Series*
(images courtesy of Waldburger Wouters Brussels and artist)

Leaf Myczack

A HOLISTIC APPROACH TO SOIL

We owe our existence to a thin layer of topsoil and rain. Given this valuable information, one would think our species would honor and cherish soil instead of treating it like dirt. We now inhabit a planet whose topsoil is covered over, eroded, degraded, overworked, and poisoned. We are long overdue for a paradigm shift about the standing of soil.

I've been on this wonderful planet for more than three quarters of a century, from a time when small family farms were the norm, animal manure was the fertilizer, and draft animals did the heavy pulling. The biologically prosperous farms were the ones that practiced an earth-based ecology that was both beneficial to the land and to the species calling that land home.

Our approach at Broadened Horizons Teaching Farm to soil restoration is based on that earth-based ecology, which is analogous to the Seventh Generation Principle. Agricultural activities today often have unforeseen long- term negative consequences. It's important to view the physical farmland in the most biologically holistic manner so that future generations can have a healthy and wholesome ecosystem. We raise dairy cows on our farm using holistic land management practices. Herd size is determined by conditions on the land, not by economic pressures. It is a blend of grass and forested hillside, and the pastures are spotted with individual and clumps of trees of various species. Wildlife moves freely in this mixed approach to open grazing, adding to the overall viability of this ecosystem to restore itself. On our larger acreage, we rest the land in the non-growing season. Our cows are kept in a "sacrifice area" and fed hay for the duration of a five-month winter. Their manure is removed daily to fertilize pastures and paddocks, and the bedding material they've urinated on is spread on the food crop areas throughout the year.

Every land use decision is approached from the standpoint of the earth's short- and long-term biological health. This involves a conscious union with the biological way of the earth so we don't trample on other ecosystems in pursuit of increased production capacity. Always at the forefront of our planning is this question: If we left this land today, could it readily convert back to a healthy segment of the southern Appalachian temperate rainforest? If not, why not? We are surrounded by farms where the answer is a resounding "no."

In agricultural societies that have farmed in place for more than a thousand years, the universal common thread is plot size.

Cielo Sand Hodson
Bounty from the Mother Earth

Very small plots, measuring two to three square meters, give the farmer an ability to know every centimeter of soil intimately. Plants reveal the health condition of the soil, and these plants also help build the soil. There are no long-term "quick fixes" to weak soil. Our small plots can get all the needed attention they require to renew themselves annually. Our best yields result while we are kneeling on the earth planting or harvesting. Humans can merge energetically with their environment to enhance positive biological results.

If an area is suitable for cultivation, or pasture management, then how much pressure or impact can we place here? And lastly, we want to biologically mitigate some of the unnatural changes we created in maintaining pastures and fields. This follows the Indigenous principle of give, take, and give back, so that there is more fertility rather than less when we are finished. Native peoples in the eastern mountains of Appalachia lived in fixed

Phil Ross *Splash*

farming villages, so renewing the soil fertility was an important part of their survival process. Indigenous peoples are centered on a sense of place, their "only home."

The land did not lose its fertility in a short period of time, and it will not be healed overnight. So to be successful with rebuilding soil, it will need to be a long-term labor of love. There are no shortcuts—biology doesn't work that way. Both soil testing and adding quick amendments and inputs that are based on human-based ecological thinking are the wrong approaches if the effort is to be sustainable in the long-term. When using imported inputs such as fish meal or rock powders, that means another part of a bioregion or ecosystem is being degraded in order to boost one's own fertility. That approach is linear thinking and zero-sum farming.

All agricultural land could use a good meal. The soil is a living organism, the sum total of microbes, bacteria, fungi, insects, root systems, and more. The healthier it is, the hungrier it gets, as life begets more life. Feed it by not viewing any contaminate-free organic material as waste, but as food for soil. Don't waste time building compost piles, spread it right on the land, the way a forest constantly adds to the very top layer of forest litter. And most importantly, treat the soil like a cherished grandmother. Talk to it, thank it, honor it, touch it, taste it, smell it; for in reality, it is you made of sunlight and soil that you are nurturing. And please, stop calling it dirt! ○

Preparations and Returns

Fire season,

kinship,

queer land stewardship,

language preservation,

creative conservation

Leke Hutchins

QUEER MAHI ʻAI CULTIVATE PUʻUHŌNUA'S IN A HETERONORMATIVE LANDSCAPE[1]

Queer mahi ʻai (farmers) growing their identity in an assertively heteronormative world have few outlets for expression. Birth families are not often accepting of a queer identity. School and professional settings can be even worse. Where is there to turn? ʻĀina (land; farm).

ʻĀina is often defined as an entity that nourishes Kānaka ʻŌiwi (Indigenous Hawaiians) physically through its bounty. Indeed, queer mahi ʻai produce thousands of pounds of food for the community every year. However, ʻāina nourishes the spirit too. In the ʻāina, queer mahi ʻai can find a reflection of themselves, a diversity-embracing entity rid of conformity which is also recovering from the scars of colonization and dispossession. Therefore, when mahi ʻai tend to ʻāina, there is reciprocal restoration occurring. While farming ʻāina, mahi ʻai can connect to akua (gods) and kūpuna (elders). They can receive guidance from revered figures from past generations, who expressed a wide range of intimacies rooted in ʻāina, of which the term queer can capture a small part of.[2] Therefore, queer mahi ʻai of today continue to lead and perpetuate Kānaka ʻŌiwi (referred to as kānaka hereafter), intimacies rooted in ʻāina.

For queer mahi ʻai, the family they choose includes ʻāina and the ʻāina's kin, both the more-than-human world of cultivated crops and animals along with the other farmers and community members who shape and share ʻāina. In hāloa, they find an accepting brother that will ensure their health.[3] They feel a warmth and embrace from the lepo (mud) of the lo ʻi (taro patch) around their legs, and, in the deepest of lo ʻi, waists as well. The many kānaka that come to volunteer and work in these spaces are kin that embrace and welcome a multitude of identities. On farms mahi ʻai and kānaka learn various forms of intimacy—such as touch or embrace—and relationship building—such as brotherhood, sisterhood, and friendship. The formation of these intimacies ties them to that space through time, creating a responsibility to continue to mālama (care for) each other.

Mahi ʻai can embody masculine and feminine traits while farming. The full range of traits can blend on ʻāina and break through the binary. When it comes to farming, a whole slew of never-ending tasks must be completed: weeding, soil prepping, planting, harvesting, and, in the context of several forms of kānaka agriculture, moving large pōhaku (stones). A heavy pōhaku that needs to be moved does not necessarily care about your sexual or gender identity. The pōhaku, along with the lepo, wai

Nanea Lum *Loli i ka ʻūmalu*

This artwork started with a land-based process—documenting the shadow of a tree which I observed over the course of a year. The shadow cast by a Plumeria tree was a place I brought my canvas in the daytime to capture its gesture. By bringing myself to this tree, the form and magnetism of the shadow created a deep connection to time and formlessness. During these long months of documentation of this tree, I began a collective project with Kanaka artist Kalikopuanoheaokalani Aiu. As a transgender performance artist, Kaliko's body is the concept of change that I observed in the dynamic variation of shape and space inside of shadows. This work focuses on the body as an ideological site of ʻāina; a place where interconnected parts compose expressions seen and felt all at once.

(water), mea kanu (crops), and other more-than-human kin are just happy to be in relation. Moreover, human kin are more concerned with your willingness to contribute and complete a task than your identity. In this way, ʻāina becomes a neutral space where binary roles are put aside.

The growth cycles of mea kanu mimic the life stages and experiences of queer mahi ʻai. The connection between the two can commemorate a momentous or joyous moment; help heal grief, loss, or pain; or create guidance and calm during new circumstances. For example, a mother can plant the ʻiewe (placenta) of their newborn child under an 'ulu (breadfruit) tree. They will then tend to the 'ulu tree and child so they may grow big and strong together. Planting can also coincide with the loss of or inability to have a child.

Cultivating a long-lasting crop can nurture you through life by providing food, shade, and comfort into your elderly years. Transgender kānaka undertaking gender affirmation can take comfort in seeing a landscape under transition due to their cultivation practices, with mea kanu displaying all forms of their stages from a planted cutting to flowering to bearing fruit.

The 'āina shaped and tended by queer mahi 'ai are pu 'uhōnua (refuge) for other queer folks traversing a heteronormative world. This queer pu'uhōnua is like how a loko i 'a (fishpond; Hawaiian aquaculture) takes in pua (young fish) and raises them in a nuturing, nutrient-rich environment where they thrive and become momona (fat). Queer kids and adults are drawn to and seek out 'āina spaces cultivated by queer mahi 'ai. The mahi 'ai sets the terms of their 'āina. This is simply more than the physical space, but the warmth and friendliness of it as well, which allows all queer folks to truly express themselves, grow in their identity, learn farming and cultural practices, and be ready to take on the outside world.

As we make adjustments to our current food system in the era of COVID-19 and social justice reckonings, we must emphasize a more equitable and inclusive system that recognizes the incredible role queer mahi 'ai play in feeding the community while raising a new generation of queer leaders. Queer mahi 'ai are leaders on 'āina and should be included on state and federal level agricultural boards and committees. From these positions, they can craft policies and rules to help fellow queer mahi 'ai based on their own experiences. For example, more grant programs should support Indigenous crop cultivation since this would support the queer mahi 'ai that primarily reside in this space, while also feeding Indigenous community members and promoting the tremendous social benefits outlined in this piece. Moreover, all queer people should feel accepted in farming spaces, no matter the identity of the farmer. This will require non-queer mahi 'ai to rethink and reset the terms of their farm, how they assign roles, and the ways in which they name, refer, and relate to people, crops, and land. ○

1. This written text was inspired by wonderful topical conversations with Brooke Leilani Hutchins, Liesel Santimer, Amber Arthur, and Kaile Luga, along with my personal reflections as a queer kānaka in the 'āina space.

2. Jamaica Heolimeleikalani Osorio, *Remembering Our Intimacies: Mo 'olelo, Aloha 'Āina, and Ea* (Minneapolis: University of Minnesota Press, 2021).

3. Hāloa, short for Hāloanakalaukapalili, was the stillborn child of Ho'ohōkūkalani and Wākea (Sky Father), who was planted into the earth and grew into the first kalo (taro). Ho'ohōkūkalani and Wākea had another child after Hāloa. That child was the first kānaka. Therefore, genealogically, kalo is considered an older brother to all Kanaka Maoli.

Luz Paczka and Olivier Herlin

NATURE'S LANGUAGE

A Manifesto

Echoes of words and voices in another mother tongue resonate through our bodies. Underneath the soil and from all of nature's sources, Earth's languages inundate our senses. But as we age, we can lose the ability to hear the phonetic subtleties across unfamiliar languages dissolving into the white noise of our surroundings.[1] All expressions of nature have their own language to be recognised and understood and, if we listen carefully enough, we might notice that nature has always communicated meaning, wisdom, and knowledge through all of our senses.

Nature's language permeates our senses. Cloud formations, tidal currents, sounds of organisms underneath leaves, the odors of ants, and each seed that exists all have a language that travels through our material world and through us. They are felt in ephemeral moments across the rhythmic cycles of the seasons, but only to the conscious observer. Imagine how we could empower biodiversity if we engaged in the unique language used by all species. These languages relay information of how each organism exists and connects to one another through diverse ways of living. But how complete is our understanding of nature if we reside within limited languages and worldviews ourselves? Do we understand nature if we do not speak nature's languages?

The Power of the Words Within

Everyday the intersection of our language and culture guides our thoughts and actions. The words we use influence the way we perceive our relationships and, therefore, our so-called reality. Descartes says, "I think, therefore I am;" an individualist perspective centered around the ego. But in Xhosa language, the word "Ubuntu" speaks a different perception of reality: "I am because you are".[2] This notion is created by a language and culture that interact to create a different way of existence: one that is not centered in an individual's perspective, but in the interpersonal relationships that form part of our existence. The interconnectedness of our surroundings can be lived, but we must learn the meaning of the words embedded in the multitude of interactions happening around us.

Adopting a language can shape how we perceive not only other people, but also other beings. About 70 percent of the English vocabulary are nouns, whereas other languages, such as Potawatomi, are composed in majority of verbs.[3] Some languages like Nuu-chah-nulth do not differentiate between noun and verb categories.[4] By prioritizing nouns in language, we influence our perception of nature. Living organisms like trees can be seen as inanimate nouns separate from their surrounding environment or seen as living verbs connected through relationships with their surrounding environment.[5]

Even if you imagine trees as living, do you see them as dynamic beings in a state of constant change? Consider their animate aspects: the abundance of ripe seeds that etch the wind and ground, the swelling and drooping from their circadian rhythms, or the transformation of trees into their afterlife of animal housing, tools, and humus. Trees continually dance between their life-giving forces and the forces of inevitable death and decay. These dynamic interactions are possible within some people's imaginations, but the collective understanding of nature in English within our discourse today has deteriorated. In certain languages and cultures there is an ever present interconnectedness in daily life. This assumption is employed in the Nuu-chah-nulth language saying "Heshook-ish tsawalk," meaning "Everything is one," which assumes that nature is interconnected, unified with the spiritual, and under constant transformation. Thus, the saying and theory of Tsawalk can be employed to understand the unique context of how things, like trees are all connected and affected by everything.[6]

Common greetings and the information we embed in our conversations can create a different account of reality. Language diversity is a source of knowledge that can also provide us with cognitive abilities that may connect us in a spatio-temporal manner. Ponder the Kuuk Thaayorre language from the Thaayorre people in Australia, which uses cardinal directions for all instances of speech. For instance, instead of saying "Move your left arm," they say "Move your southeast arm" (if it was in the southeast direction). Additionally, instead of greeting people with "How are you?" the Kuuk Thaayorre ask, "Where are you going?" Therefore, Thaayorre people must always be cognizant of their present and future location in space. The way they engage in conversation explains their exceedingly high ability for spatial cognition.[7]

What would the world look like if we directed our cultural and linguistic discourse to support reciprocal relationships with our surroundings, and specifically the natural world? What other influences do we want from our cultural discourses and the way our languages shape our thoughts, actions, and worldviews?

Can We Listen Carefully Enough to Hear Nature's Language?

Unfortunately the degree of listening and processing required to learn a new language is quite demanding and goes beyond literal translations. With an internet search, we may think we understand what "Mní wičhóni" (water is life) and "Miigwetch" (thank you) mean; from an article on non-human languages, we may think we know the message and meaning behind interpreting animal tracks, the scent of ant pheromones, the presence of a bioindicator, or invasive species.[8,9] But think again. Imagine translating the word "sustainability" into a language of consumer capitalism. Grasping a concept from a new language requires more thought than the assimilation of words into your own worldview. There is a difference between changing a word's meaning to accommodate the interpreter, and changing worldviews to accommodate the real essence of a word. While translations may lead us towards new mindsets and languages, they can also lead us astray.

Similar but distinct from a translation is the power of story. By listening to a story with empathy and imagination, we immerse ourselves in another being's experience beyond our own ego and individuality. By experiencing a story, we can begin to understand the underlying lesson behind the constantly evolving oral history of the earth. Stories have profound influence on the multitude of truths each person experiences, and their role in the collective culture, wisdom, and knowledge.[10] However, these oral stories are often tied to language and the traditional meanings that are the essence of these lessons. To become connected and more aware of the natural world around us, we must preserve languages and stories to shape how our collective discourse evolves in connection with nature.

The Present and Future Power of Language

Stories from language travel like sounds, echoing across valleys and permeating through the thickest of walls. Their evolution and distribution cannot be stopped. Echoing through our minds, there are stories that show us how to connect and interact with nature. The story of quelite confronts the fear of using foraged foods by showing evident human connection is an example.

This story lived by many begins with a question: "Is that edible?" A person points to a plant among the desolate urban forest. In

answering, we attempt to share the empowering qualities of foraged wild edible plants. However, English brings into question the quality of our foraged choices, because "edible" posits another question: "How edible?" Then we are left with a futile task of convincing and changing a person's belief about whether wild plants are appropriate for consumption. Consider instead the word "quelites," a Nahuatl word for "nutritious, edible, and local plants, both wild and cultivated."[11] This English translation falls short of the word's real meaning, which brings agency and purpose to the unexpected presence of plants. The ancestral knowledge of Nahuatl has been passed on for generations and, even now, about 350 quelites are sold and eaten in Mexico without question, including purslane, papalo, lamb's quarter, and mallow. Quelites is a word that speaks a form of nature's language, evolving and adapting into the present human discourse to let wild plants coexist among humans. It shows the strength of using nature's language in innovative and creative minds to turn current discourses into ones that care for, and listen to nature.

Nature's languages have persisted through war, colonization, humanity's darkest hours, and a changing climate. We can remember them with our ancestors, learn them, and continue to speak words of power. We can learn the languages of the birds, clouds, insects, leaves, soils, seeds, and waters, so we can journey together on the path of becoming rooted to place.

Learning our ancestral languages taps the ancestral knowledge within us all, but we can never be rooted to place in a persistent colonial culture. We will always be displaced into a nomadic lifestyle as a result of injustices of our own design, traveling and working on lands and waters we will never be able to own. Nonetheless, on our continual migration, we can share the seeds of nature's language and the continuance of an oral history and way of knowing in the world. We can learn these words to see the world as it is now and use them as a tool for resilience that will help us collaborate when we join forces during challenging situations. Recognising and speaking nature's language can empower our communities as we learn about the persistence and resilience of nature. Each language of the world brings new ways of understanding nature and experiencing the world, and can shift our discourses to new ways of thinking and living. ○

Felipe Delfino *Kiva*

Felipe Delfino *Untitled*

Notes

1. Patricia Kuhl, "The Linguistic Genius of Babies," *TEDxRainer*, February 2011.
2. Mungi Ngomane, *Everyday Ubuntu: Living Better Together, the African Way* (New York: HarperCollins, 2020).
3. Robin Wall Kimmerer, "Learning the Grammar of Animacy," *Anthropology of Consciousness* 28, no. 2 (September 2017): 128–34.
4. Jan Rijkhoff, "When Can a Language Have Nouns and Verbs?," *Acta Linguistica Hafniensia* 35, no. 1 (January 2003): 7–38.
5. Dedre Gentner, "Why Verbs are Hard to Learn," in Roberta Michnick Golinkoff and Kathy Hirsh-Pasek, eds., *Action Meets Word: How Children Learn Verbs* (Oxford: Oxford University Press, 2006): 544–545.
6. Umeek (E. Richard Atleo), *Tsawalk: A Nuu-chah-nulth Worldview* (Vancouver: UBC Press, 2007). 117–131.
7. Lera Boroditsky, "How Does Language Shape the Way We Think?," in Max Brockman, ed., *What's Next* (New York: Vintage Books, 2009). 116–29.
8. James Vukelich, *How to Say "Thank You" in Ojibwe! Ojibwe Word of the Day Miigwech ᒦᒀ "Thank You,"* October 21, 2019. YouTube video, 12:36.
9. Tao Orion, *Beyond the War on Invasive Species: A Permaculture Approach to Ecosystem Restoration* (White River Junction: Chelsea Green Publishing, 2015): 166–170; 239–243.
10. Basil H. Johnston, *"Is That All There Is? Tribal Literature,"* in Jill Doerfler, Heidi Kiiwetinepinesiik Stark, and Niigaanwewidam James Sinclair, eds., *Centering Anishinaabeg Studies: Understanding the World Through Stories* (Winnipeg: University of Manitoba Press, 2013): 3–12.
11. Mahelet Lozada Aranda and Lucila Neyra, "Quelites," Biodiversidad Mexicana, Comisión Nacional para el Conocimiento y Uso de la Biodiversidad, updated May 19, 2021, *biodiversidad.gob.mx/diversidad/alimentos/quelites*.

Sharifa Oppenheimer

FIRE AND LIGHT

Fire's incandescence dies
into silk. Albedo. Ash.
Cool, nutritious
alabastrine as spider's web
it sweetens acid soil
warms with a tender hand:
      destruction eclipsed
      by grace.

Ash, nutrient rich,
dug back into human soil
alkalizes
phosphorizes
bio-luminesces.
No longer making
light by heat
we metamorphose.
Now
      our lucid prayers
      and conversations
      with glowing fungi

are coronas of light
in the dazzling dark:
      we become
      the communion
      of light upon light.

A version of this profile with the same title was initially published in A Litany of Wild Graces: Meditations on Sacred Ecology *(Red Elixir Press, 2022), and was printed by permission of the author.* ○

anna ialeggio

THE SECOND FIELD

In her 1986 essay "The Carrier Bag Theory of Fiction," Ursula K. Le Guin reflected on the role of ecologically oriented and feminist science fiction: "Lest there be no more telling of stories at all, some of us out here in the wild oats, amid the alien corn, think we'd better start telling another one, which maybe people can go on with when the old one's finished."[1]

I grew up incessantly reading Le Guin's books. Each one is like standing on an iceberg. I can feel the enormous mass of it below and around me, complex and integral to its own logic and poetry. It became apparent to me over time that each had been written to ask a question. What is gender? What might it look like to live outside of capitalism? How does time work? Can dislocated relationships with land be healed, is it possible to live in accord with the more-than-human world?

These questions are never answered didactically, but are a place for reflection that author and reader return to, again and again, with the evolving tools for interpretation that we gather throughout our lives. The changing discursive ground where I, the reader, stand in relation to a book is central to the experience: over time I arrive with different questions and leave with different answers. It is the kind of reflection that challenges my understanding of

self—who I am and what I might be capable of doing or making in this life.

I've recently returned to Le Guin's work with prairie restoration in mind. Prairie restoration seeks to establish and maintain precontact ecologies and plant communities against the advance of vigorous opportunist grass species, many of which were introduced in North America for, or as a direct result of, agricultural purposes.[2] The opportunists outpace the ability of an ancestral prairie to thrive without intervention. Among other characteristics, they tend to form thick rhizomatic root mats through which the prairie plants have difficulty penetrating. It is often challenging not to imbue the troubled relationship between so-called native and invasive plants with the values and habits of human beings, and to see the postcontact species as metaphorically, as well as ecologically, harmful.

Marine fisheries scientist David Pauly said the following:

> We transform the world, but we don't remember it. We adjust our baseline to the new level, and we don't recall what was there... Every generation will use the images that they got at the beginning of their con-

anna ialeggio *Fire at Pachamanka*

scious lives as a standard and will extrapolate forward. And the difference then, [we] perceive as a loss. But [we] don't perceive what happened before as a loss.[3]

Pauly coined the term "shifting baseline syndrome" in 1995 to describe the limitations of using human perceptions of change to inform environmental conservation policy-making or management.[4] The term—scientifically precise and philosophically elusive—suggests that our ethics are the result of slippery, simulatory concepts, appearing and receding within human perception at great speed, rather than solid or empirical. In this light it appears that empiricism itself is not solid, and that what we strive to conserve or rebuild depends very much on what we perceive to be the baseline for a healthy ecosystem. Pauly and his

colleagues often use the example of conservation focused on pasture restoration in the English countryside, which often produces homogenous landscapes because the desired baseline is derived from the historic effects of intensive sheep grazing after mass deforestation.[5] This pastoral "viewshed" is an aesthetic affirmation of human ideas, practices, and real estate values—not a restoration of biodiverse grassland, oriented to the needs of its other-than-human inhabitants.

Pasture restoration as an aesthetic rather than ecological imperative, derived from the concentration of wealth by elite landowners, is connected to the expectation that a house ought to have a lawn, an artificial pasture.[6] A lawn is stable only through endless nudges and agreement to conform to a certain aesthetic that in turn, confirms participation in social norms: paying a teenager to mow, applying fertilizer and herbicide from big box stores, and neighbors and condominium associations complaining about dandelions. In the part of upstate New York where I live, the lawn can reach out from the house until it becomes a pasture: nonfarming landowners who would like to retain a lower tax status for land in current agricultural use will frequently lease to a farmer for hay because that process produces

anna ialeggio   *Rainbow over the gardens*

a nice-looking field. It is a tidy form of agriculture without any mess or animals.

For me, the idea of shifting baselines expands beyond the ecological to include creative legacies. In both realms, we make decisions about what to do, make, and think based on our perceptions of what is "normal" and "natural" to do, make, and think. When I apply the metaphor of pasture restoration, for example, to my own practice of making artwork, the links between my own habits of imagination and aesthetic conformation come into view. A momentary slippage in perspective like this opens up room to poke and prod at those norms: I can make adjustments, observe changes over time, and then make further adjustments.

Adjusting a baseline sounds simple but in practice it is far messier. I think of Pachamanka: a fifty-acre piece of land stewarded by Nance Klehm in the Driftless Area of Northern Illinois, Ho-Chunk territory. There are mixed woods and fields—agricultural land in transition from decades of hard use to something else. On the northeast field, there are several acres of reconstructed prairie that previously grew corn, soy, and hay. It was plowed and seeded

with native prairie species. The opportunistic grass that vexes this area is reed canary grass *(Phalaris arundinacea)*. Nance's prairie is constantly encroached by Phalaris coming in from the farms and fields around; every spring, the prairie needs to be burned and resown by hand with seeds that were gathered in the fall. Reclamation is not a single act, but an ongoing and cyclical process—as well as an artificial one.

I attended the prairie burn at Nance's on a spring day when the winds were low. The fire crew worked their way across the meadow, the flames creeping through the dry tinder on the ground ahead of them. I watched the smoke blow through the trees from the woods. Later that week, on a day heavy with rain, Nance and I went out and slung seeds mixed with sand through our fingers, casting them over the burned prairie. This spring I went back out to the Driftless and we picked our way through bergamot and goldenrod, bending to touch the new growth.

In mathematics, perturbation is the complex motion of a massive body subject to forces other than the gravitational attraction of another massive body.[7] The motion that the earth and moon follow, for example, under the gravitational effect of only one on the other would be considered geometrically simple. The effect of the Sun upon this stately two-body dance creates complexity and would be considered a perturbing force; nudging things out of their expected motion. The entire system produces nudges as a collection of forces, processes, and reactions. The agency of the nudge itself is inseparable from the distant observer.

Perturbation theory is a useful way of describing the various dynamics at play in Nance's prairie. As I walk through it, the prairie nudges me back in the most unpredictable

→ **anna ialeggio**   *Prairie burning at Pachamanka*

ways, asking me to consider how I interact with various legacies of caring for, or using, land. It reminds me of Le Guin's books, which examine and abrade the default settings of Western objectivity through a stubborn, curious humanism—rather than celebrating or articulating this world as it appears to be, the author's fictional worlds pull up the monoculture species, return something to the soil, and slowly make room for more to grow. There is particular resonance here because at Pachamanka, there is a second field.

The second field is less clearly defined than the reseeded prairie, extending from the woodlot—walnuts and pines planted on a square grid by a previous landowner—to the creek and up the hill to the west where it adjoins a neighbor's cow pasture. This field wasn't plowed or reseeded and is on its path to become what it will be next. Thickets of box elder spring up and coyotes trace through. The deer voraciously eat anything that is planted. Sandhill cranes summer here, and white pelicans stop by a seasonal pond in spring, smaller and drier every year. The giant old white oaks by the creek seem to have developed a fungal infection, and the ends of their branches are turning brown.

As evocative as the first field may be, with its ongoing, devoted reclamation and maintenance, Nance says that the second field is where we are all at. The legacy that we've collectively inherited is a vast, disturbed,

and transitional field. Understanding this, we can observe, learn, and occasionally make or receive a tiny nudge to see if there are ways that we can be helpful. But above all, this field asks us to commit to the complexity of a less than picture-perfect narrative about ecological reclamation and healing, to go out into the alien corn and wild oats. We can return to the wild grasses, to the experience of actually standing in a field on a sunny day, in the awareness that what may look uniform is in fact an lively ongoing exchange in a time of colossal adjustments. Lest the stories, lest the prairies, end up sounding and looking like only one story, like only one species of grass. ○

Notes

1. Ursula K. Le Guin, *Dancing at the Edge of the World: Thoughts on Words, Women, and Places*, (New York: Grove Press, 1989), 165–170.

2. Martha J. Groom, Gary K. Meffe, and C. Ronald Carrol, *Principles of Conservation Biology*, (Sunderland: Sinauer Associates, 2006), 314–316.

3. Daniel Pauly, "The Ocean's Shifting Baseline." Filmed February 2012 at Mission Blue Voyage. TED video, 8:46.

4. Daniel Pauly, "Anecdotes and the Shifting Base-Line Syndrome of Fisheries," Trends in Ecology and Evolution 10, no. 10 (October 1995): 430.

5. Allison Guy, "Daniel Pauly and George Monbiot in Conversation About 'Shifting Baselines Syndrome,'" *Oceana* (blog), August 14, 2017.

6. It has been frequently noted that one of the earliest modern lawns include the 17th–18th century Versailles gardens of Louis XIV of France. These pleasure gardens included squares of *tapis vert*, or "green carpet," created to resemble heavily browsed sheep pastures, though often mowed by human servants.

7. Richard Fitzpatrick, An Introduction to Celestial Mechanics, *Cambridge: Cambridge University Press*, 2012, 149.

Rose Robinson

OH, SWEET ONE

Who is going to tell this little ash sapling
how hard life is going to be.
It just came forth from the soil and
so soon it will return with its momma,
as the mushrooms grow lively
with the stink of death. Sweet one,
I wish I could sustain you, but I am unable
to home a tree and shield off the wild world.
I too fear the borers and the fires.
We are all at the mercy of the sun.

Vincent Waring   *Twisting Stories*

Ethan Young

THE ROLE OF ECOSYSTEM FUNCTION IN CATASTROPHIC FIRE PREVENTION

Introduction

Ecosystem function is not often discussed with regards to planning, policy, law, and strategies to address many issues threatening the security and welfare of our species, even though it is fundamental to those issues. The following discussion will focus primarily on how we can positively impact ecosystem function with regard to the intensity, frequency, and duration of catastrophic fire.

Defining the Problem

We as a species are trying to prevent catastrophic fires, not wildfire generally. As a sub-type of wildfire, catastrophic fires burn with sufficient frequency, intensity, and duration to cause major disruptions in ecosystem function, drastically set back ecological succession, and endanger habitat for countless species, including but not limited to humans. Catastrophic fire is, at this time, mostly the result of chronic human mismanagement of terrestrial landscapes and resources. We have, through our management decisions and actions, created the perfect conditions for the normalization of wildfires of catastrophic intensity, frequency, and duration.

A toxic buildup of chemical energy occurs when biological (metabolic) energy flows and nutrient cycles are disrupted, diverted, shifting the emphasis to abiotic (chemical) processes such as oxidation and combustion. By shifting the emphasis back to biological processes, we can sequester carbon, limit fuel buildup and reverse the climatological factors contributing to catastrophic fire.

Where Is the Water?

Biological processes require humidity. A few hundred miles from coastal climates, a very large portion of inland precipitation occurs through a process known as bioprecipitation, as opposed to oceanic evaporation. Bioprecipitation occurs through mass respiration, most of which involves evapotranspiration from plants as they undergo photosynthesis. As the land breathes, it ejects massive thermal currents of moisture into the atmosphere along with specific microorganisms responsible for condensing water vapor.[4] A living landscape full of diverse biological activity is breathing, and in breathing creates its own precipitation. For this reason, I call bioprecipitation the "bucket brigade," because it's how ecosystems carry and recycle freshwater inland to replenish terrestrial watersheds on a seasonal basis. The more diverse and active an ecosystem, the more it contributes to bioprecipitation, and the more reliable its water source becomes. This is especially true with the restoration of the soil carbon sponge, which absorbs, stores, and steadily releases water throughout the season. An intact soil carbon sponge produces a regulatory effect on water supply, improving its reliability which further enhances the ecosystem function by removing water as a limiting factor for photosynthesis and related biological activity.

Drought slows the biological decay of organic matter, and optimizes the ability of fine fuels to burn hot and fast. Runoff/erosion/flooding and drought are two sides of the same coin, symptoms of an impoverished water cycle. The water cycle can improve through maintaining a good ground cover—both living herbaceous plants and plant litter—and sequestering carbon through the liquid carbon pathway. An improved water cycle enhances the ability of the soil carbon sponge to absorb, hold, and steadily release an accumulation of water over the course of a season, recharging groundwater and reactivating perennial springs.

An ecosystem with improved water sequestration and holding capacity will stay green and active longer, and will burn less readily. Nutrients will cycle and energy will flow biologically for more of the year, leaving less available for fire to consume. By rejuvenating the carbon soil sponge, watersheds become a more reliable source of clean, fresh water for everyone.

The living ground cover also keeps both the air and soil itself cooler while providing a source of humidity for the air, which increases the frequency, intensity, and duration of dew point and reduces evaporation from the soil.

A Tale of Two Carbons

There are at least two dominant carbon pathways in an ecosystem. Lignification describes photosynthetic activity that leads to a buildup of carbohydrate biomass comprising the physical structure of plants. This pathway provides an important contribution to the floral biomass that increases overall photosynthetic capacity and feeds fauna, but it is also the primary way that fuel buildup occurs if left imbalanced with the other dominant carbon pathway. Overemphasis on lignification can smother many herbaceous perennial plants and cause an ecosystem to reach a decadent closed-canopy climax state resulting in a decline in net primary productivity and diversity compared to mid-succession ecosystems managed through intermediate disturbance.

In contrast to lignification, the liquid carbon pathway (LCP) is where plants synthesize and deliver sugars down into the soil food web through their roots and symbiotic connections with fungal hyphae to feed the soil biology, converting sunlight and CO_2 to sugars and sugars to stable humus.[5] The LCP is generally much more stable as a form of carbon sequestration and, over time, it turns dead mineral substrates (a.k.a. dirt) into a living organism built increasingly out of the dead bodies and exudates of a vigorous soil food web (a.k.a. soil) that in turn renders important nutrients bioavailable to both plants and the above ground animals that eat them.

Animal Impact to the Rescue

One of the primary ways to prioritize the LCP over lignification is through intermediate disturbance.[6] Grazing and related forms of animal impact represent the primary tool through which we can provide solar-powered biological disturbance to an ecosystem to enhance function. Grazing utilizes plant biomass and prevents lignified fuels from accumulating. When implemented according to the principles of trophic cascade ecology (i.e., the predator-prey dynamic)—through planned grazing, for example—it also optimizes energy flow and nutrient cycling over a growing season, improving ecosystem structure and diversity and increasing photosynthetic activity in general.[7]

Because of the enhanced nutrient cycling, that biomass doesn't accumulate as lignin and fuel load. Instead, it gets redistributed to organisms both above and below ground through metabolic activity in the food web as dung and urine, and through scavenging, predation, and decomposition. This biological nutrient cycling translates into increased biodiversity and biological activity. However, we often use abiotic forms of disturbance— such as fire, chemicals, mechanization, unplanned grazing, or over-rest—that degrade ecosystem function over time. As complex systems, ecosystems don't just recover from disturbance. They undergo a process of adaptive recovery. If fire is the dominant form of disturbance in an ecosystem, then the ecosystem's recovery process will start favoring the dominance of fire-dependent and fire-tolerant vegetation. By balancing forms of disturbance, and specifically prioritizing biological and metabolic forms of disturbance over chemical and mechanical disturbance, we allow greater diversity amongst primary producers.

A well-grazed ecosystem may have less above ground biomass at any given point than one that is rested due to enhanced biological decay and localized nutrient cycling. However, under the same conditions it produces more biomass overall throughout the growing season and distributes that biomass as nutrients more effectively to provide habitat for a greater number and diversity of consumer species. We can think of above and below ground carbon like an iceberg. Lignification is like an inverted, top-heavy iceberg. In contrast, a liquid carbon pathway in balance with lignification will look the way an iceberg should, with only a relatively small, though still abundant, amount of above ground biomass, and a relatively large below ground mass of biological activity in the soil food web. While in any single snapshot, the above ground biomass of a lignification-focused ecosystem might look more impressive than an ecosystem focused on the LCP, the better measure of biological activity is through measuring the total amount of consumer biomass, and the lengths of food chains that make up the complexity of food webs in the ecosystem over time. In this way, well-managed grassland or woodland ecosystems can be incredibly—even deceptively—productive and diverse to the untrained eye.

Another important form of animal impact is the trampling of biomass into litter cover. Litter armors the soil against UV radiation and extreme temperatures, prevents evaporation, compaction, and erosion. Litter facilitates water sequestration by slowing and spreading the flow of water in conjunction with the creation of countless little random microcatchments that optimize soil habitat for flora and fauna and improve seed germination.[8]

Alongside grazing, trampling turns the liability of fine fuel loads into an ecological asset. Trampled and grazed biomass holds more moisture and becomes more bioavailable as fuel for organisms rather than chemical fuel for fire.[9] These "soft" (biotic) firebreaks are almost as effective as "hard" (abiotic) firebreaks and contribute much more to ecosystem function than chemically oxidizing lignified biomass or denuded sites.[10]

None of this is to say that fire is intrinsically bad. It is a way that nutrients cycle and an important intermittent source of disturbance and renewal. But it is a large, nonlocal nutrient cycle, rather than a small, local nutrient cycle. Likewise, fire is one of the most intense forms of disturbance an ecosystem can undergo, short of the scorched-earth chemical warfare we call industrial agriculture. The more intense the fire—or disturbance, generally—the less frequently it should occur for optimal ecosystem function.

Relatively small, cold fires that leave at least some groundcover and soil life intact can contribute to intermediate disturbance and renewal to keep ecosystems functional and ecological communities in balance. It is arguable that even these small, cold fires should not generally occur every year. But right now, we see the equivalent of catastrophic century— or millennia—fires on an unprecedented, almost annual, frequency. This sort of frequency, intensity, and duration of catastrophic fire disturbance is one of the final nails in the coffin of anthropogenic desertification, or the human conversion of ecosystems into desert wastelands. Like grazing, we've been using and interacting with fire poorly. By optimizing grazing management and resulting animal impact to prioritize ecosystem function, we can also optimize the role of fire and begin to minimize the combined threats of climate change, catastrophic fire, flood, drought and erosion.

Some of this ecosystem function can be designed and engineered. For example, Yeoman's Keyline planning and design according to the Scale of Permanence can improve the water cycle and contribute to the rehydration and functional beautification of both rural and urban landscapes.[11, 12] But ultimately, community dynamics—the interaction between organisms, populations, and species in an ecosystem, generally—and trophic cascade ecology, are key to integrating humans into the optimal management of ecosystem function over time. This means understanding the role and behaviors of ethical predators in an ecosystem, in part to respect and support the matrilineal structure and function of herbivorous herds of megafauna. By appropriately prioritizing and valuing ecosystem processes and ecological function, we can tackle many of the environmental, ecological, and climatological crises we face simultaneously.

Valuing Spaceship Earth

A recent article asked why Californians don't use grazing animals more widely to reduce fire risk, compared to mechanical and chemical fuel load reduction, but failed to discuss any-thing beyond the "high and rising cost" of ecosystem service grazing, including its potential value beyond mere fuel load reduction.[13] Neglected measures of value in ecosystem service grazing include lives and infrastructure saved, drought and floods mitigated, and ecosystem services provided in the form of biodiversity, carbon sequestration, and water and air quality. Devaluing ecosystem services—and ecological function in general— is unfortunately normal in our society. It's what happens when we devalue and treat complex life processes like mechanical objects.

Function is a process-based concept, and all processes depend on relationships. If something is functioning well, it works well in reciprocal relationship to its context. Object-oriented perspectives that try to impose static ideals end up fighting ecosystem function and undermining the integrity of those relationships. No amount of caring about ecosystems as an object will regenerate ecosystem function. All too often, we only give lip service to caring. But when we care about ecosystem function on its own terms, we will allocate sufficient resources toward supporting functional processes and relationships, and the appropriate allocation of resources will provide evidence of caring.

Ecosystems are complex living, breathing, evolving, adapting systems, and we need to treat them as such. The final component of addressing catastrophic fire involves a cultural and economic shift of values among the general population to drive shifts in policy and law that support the appropriate integration of humans as managers in the broader landscape. Until this cultural, economic, and political shift occurs, we will fail to see desirable outcomes in landscape management activities, and the situation will continue to worsen. Ultimately, we need to become as committed to ecosystem function as we were to sending a man to the moon. Where there's a common will, there's a way.[14] ○

Notes

1. Anthea S. Jones, Byron B. Lamont, Meridith M. Fairbanks, and Christine M. Rafferty, "Kangaroos Avoid Eating Seedlings with or Near Others with Volatile Essential Oils," *Journal of Chemical Ecology* 29 (December 2003): 2621–2635.

2. Allison T. Karp, J. Tyler Faith, Jennifer R. Marlon, and A. Carla Staver, "Global Response of Fire Activity to Late Quaternary Grazer Extinctions," *Science* 374, no. 6571 (November 2021): 1145–1148.

3. Aristides Moustakas and Orestis Davlias, "Minimal Effect of Prescribed Burning on Fire Spread Rate and Intensity in Savanna Ecosystems," *Stochastic Environmental Risk Assessment* 35 (January 2021): 849–860.

4. Brent C. Christner, Cindy E. Morris, Christine M. Foreman, Rongman Cai, and David C. Sand, "Ubiquity of Biological Ice Nucleators in Snowfall," *Science* 319, no. 5867 (February 2008): 1214.

5. Christine Jones, "Liquid Carbon Pathway," *Australian Farm Journal* 338 (July 2008).

6. Rebecca L. Brown, Lee Ann J. Reilly, and Robert K. Peet, "Species Richness: Small Scale," *eLS* (2016).

7. A classic case study for trophic cascade ecology involves the biocidal extermination of wolves from the Yellowstone ecosystem, followed by declines in biodiversity and ecological function alongside the desiccation of the landscape; whereas the more recent reintroduction of wolves has led to the reversal of ecological decline and a resurgence in biodiversity, ecological function, and rehydration of the landscape.

8. Alejandro Carrillo, for example, regularly notes soil temperature differences of 80° and 150°F between soils covered in litter and living plants and bare soil in the Chihuahuan desert.

9. Dylan Schwilk, "Flammability Is a Niche Construction Trait: Canopy Architecture Affects Fire Intensity," *The American Naturalist* 162, no. 6 (December 2003): 725–33.

10. Soft firebreaks may gain much of their effectiveness from enhanced water cycle function resulting in higher consistent fuel and soil moisture, as well as from reduction in surface area to volume ratio of fuels and reduction of overall fuel load, whereas hard firebreaks gain their effectiveness from the near-complete elimination of carbon from the landscape.

11. The basic idea is to base landscape design planning processes on a cascading scale of permanence and importance, to "buy the tie to match the suit," so-to-speak. Darren Doherty's Regrarian's project synthesizes broadacre ecological design and holistic management using the scale of permanence as the organizing structure and process. See *smallfarms.cornell.edu/2016/04/scale-of-permanence*.

12. P.A. Yeomans, *The City Forest* (Sydney: Keyline, 1971).

13. Emma Talley, "Why Don't Californians Use Goats and Sheep More Often to Reduce Fire Risk?," *San Francisco Chronicle*, July 17, 2022. See *pubs.usgs.gov/of/2008/1214/pdf/ofr20081214.pdf* for another example of focusing on fuel load to the exclusion of ecosystem function in moderating fire spread, intensity, frequency and duration.

14. Undiscussed here is the role of Indigenous cultures in landscape management beyond fire, as predators in a regime of applied trophic cascade ecology. White control of discourse has fragmented, marginalized, or erased many important aspects of Indigenous landscape management, narrowing the story of their complex relationship with the landscapes they call home.

Liz Toohey-Wiese *Two Horses*

In June 2021, a historic heat dome lingered over most of Western North America for over a week. In British Columbia, we saw the highest ever recorded temperatures recorded in Lytton, at 49.6°C. A day after this temperature was recorded, the entire town burnt down from a fire thought to have been started by a train passing through, its brakes sparking and catching the dry grasses on fire.

A post popped into my Facebook feed about evacuating livestock during wildfire season. If an encroaching wildfire is approaching too quickly and there is no time to evacuate your animals, the best thing you can do for them is spray paint your phone number on them, open all your paddocks, let them go, and hope for the best. I thought that this seemed like a poetic image of human acceptance relating to the possibility of loss in the face of these conflagrations.

This painting was made from ash collected from the Mount Christie wildfire in Penticton, British Columbia.

Mary O'Brien

THE INDIVIDUAL IMPACTS OF ENGAGED COMMUNITIES

The bigleaf maple tree I've lived under for nearly half my life is starting to die. Arborists consider this stand of thirteen massive trunks one tree, and date it at over one hundred thirty years old. It is presumed to be the largest tree of its kind in my county. It shades our house most of the year, even a bit without its leaves in the winter. But to do this, its generous canopy leans over a third of the house. It's been regularly trimmed and examined by arborists, but its lean is scary.

My partner and I have been caring for this tree for nearly a quarter of its life. Before our one-hundred-year-old "summer shack" cabin was built in the redwoods north of the Golden Gate by San Franciscans escaping the seasonal fog, the tree was already thriving. At that time it would have been just a cluster of light green leaves set against the dark redwoods that populate our hill. I imagine the home's first residents were pleased with the tree's gift of privacy, orienting their screened-in porch to take advantage of its cover. Now, its massive, moss-coated trunks rise like columns to frame our dining room windows. Someday, when a new homeowner cuts it down, over a century of active carbon sequestering will be eradicated.

The term carbon sequestering has reached the vernacular. That we should pay more attention to our everyday carbon footprint is understood. A century of climate-compromising car emissions, non resilient farming methods, carbon-binging architectural practices, the methane from overflowing landfills, and cows—millions of cows—have taken their toll. Less known, and more perplexing, is how an individual can contribute to the enormous task of sequestering carbon in a measurable way. The path towards net-zero carbon is a narrow one. Finding a personal, virtuous, "better" way to live in our carbon-belching society waxes unattainable.

Individually, we're removed from interacting with the technology that stores large amounts of carbon through sequestration hubs. Most of us don't have the daily access necessary to significantly contribute to pulling carbon out of the air: think kelp forest regeneration, farming perennial grains, and innovative reforestation techniques. But carbon sequestering is just part of the carbon remedy; carbon stewardship is needed as well.

When we make the connection between our actions and the environment's reaction, we steward. While making day-to-day reductions in our own carbon footprint through properly utilizing farm wastes, installing wind power, and capturing landfill gas to generate electricity,

Studio of Watershed Sculpture (photography by Mary O'Brien)
Sky Opening: Riley Road Tree Plantation, Interlochen, Michigan
Overgrown and unmanaged red pine plantation trees are trimmed
to open the canopy to light, rain, and wind. This sculpture allows
the process of natural reforestation to begin while avoiding the
soil degradation and other loss of habitat due to the effects of
clear-cutting.

for example, is not as easy as popular websites promise, stewarding captured carbon can become inherent in each of us. With a fortified connection to the commons—our shared cultural and natural resources—individuals can envision their ideal restored world.

Thoughtful interventions such as re-naturing our neighborhoods with drought-tolerant plants, pollution-sucking native trees, resource-efficient rain gardens and gray-water systems, low-impact lighting, pesticide-free gardens, and animal-friendly infrastructural elements impart knowledge and build wisdom. We can discover ways to steward our own lands and eventually influence the public domain. We can engage, we can advocate, and we can demand better. When we put carbon-saving practices into our homes, backyards, and sidewalk medians, we affect a shift in discern-

ment within our communities that can facilitate protectionism. Stewardship creates agency. It allows communities to expect more from our governments, from industry, and from ourselves.

We steward by doing the work that sustains a cultural shift, advocating for more revered, protected, and reclaimed lands. Here's where artists and other creative thinkers can play a practical role. Artists have a platform for transformation. They can make activism visible and memorable. A well-researched, well-intentioned, creative person in the mix becomes an instigator to adjust perceptions, exposing sustainable consequences. Those results can be beautiful, as well. It will require imagining different pathways. It demands fortitude, and patience. Generations of patience.

Studio of Watershed Sculpture (photography by Chris Hintz)
Sky Opening: Riley Road Tree Plantation, Interlochen, Michigan

Studio of Watershed Sculpture (photography by Mary O'Brien)
Line of Defense: Mississippi River Delta, Venice, Louisiana
On land that was dissolving due to sea-level rise and related storm
surge, a disappearing island in a former cow pasture is the site
of a living sculpture. In this artist intervention, newly planted
bald cypress seedlings are protected from predation by an invasive
aquatic species, the nutria.

If the bigleaf maple tree sheltering our
house is logged someday for safety reasons,
or fire protection, or because a future resident
decides it creates too much shade, the wood
will likely be mulched. Unless it can be
harvested as lumber to build homes or furni-
ture, its one-hundred-thirty years of stored
carbon will be released quickly through
decomposition. Sure, new plants and trees
will benefit, and perhaps continuing to
maintain this tree in its old age will be sufficient stewardship to cover for the previous
five to six generations of property owners
who lived under its shade. But today, I'm going
to start looking for a contractor interested
in maple logs. ○

Connectivity

Assisted migrations,

habitat mosaics,

trickster ecologies, riverways

Renée Rhodes and Alli Maloney

PLANTING FOR FUTURES PAST

An Interview with Oliver Kellhammer

In 2002, ecological artist, educator, activist, and writer Oliver Kellhammer began planting trees in his yard on Cortes Island, British Columbia. His experiment—now a decades-long botanical intervention called *Neo Eocene*—focuses on the adaptability of prehistoric trees, types that made up the area's Eocene forests millions of years prior. To develop insight about forest adaptability and how to help the environment develop climate survival strategies, Kellhammer has continued this planting project ever since. Tracking progress along the way, he scaled up in 2008 when he began to plant a grove on sixty acres of clear-cut land.

The following conversation explores the project's trajectory from the Eocene to life beyond the Anthropocene. This interview has been edited and condensed for clarity.

What sparked your curiosity about the forest planting that you're working on?
I am an artist who has worked with landscape since the 1980s when I was living in British Columbia in this beautiful rainforest where there was a lot of logging going on. Anybody who lives in the Pacific Northwest knows about these huge clear-cuts, where these old growth forests get logged off which leave bare areas that get reforested by industry. They tend to be

monocultures. So, I was aware of that. I had been living in that area for a long time. And there's this specter of climate change: we are experiencing anthropogenic climate change where temperatures are forecast to go up four degrees Celsius, which is a lot, in a hundred years. We're already on our way to two degrees Celsius within the short term. The temperature's getting hotter, and yet we've cut all these trees down and we're replanting the forest with the same trees that were there before. I was interested in asking, is this a good idea? And as an artist who uses my imagination a lot, what would be a more interesting thing to do?

I came upon this concept of bringing back prehistoric trees, trees that had grown in the area during the last period of global warming or intense global warming, which was fifty-five million years ago. I thought if I could figure out what grew back then and bring those trees back, then maybe the landscape could adapt itself to the effect of global heating or climate change, more sustainably and less violently.

What are some of the tree species that you started planting, and how did you choose those specific species?
In the environmental movement, we're very concerned about bringing in a species from

Oliver Kellhammer
Metasequoia: 10–12 years old, 8 meters in height →

Oliver Kellhammer   *Eocene Conifers*

elsewhere to a place where it shouldn't be. The trees that we chose were prehistorically native, so they weren't never there. They just weren't there recently. Fifty-five million years ago, there weren't any human beings around, so we have to rely on the fossil record. The area of Cortes Island, where the main project is, does not have sedimentary rocks, but maybe less than a hundred miles away are areas

with extensive deposits of fossilized leaves of trees—perfectly preserved fossils of California redwood, like coast redwood, sequoia, dawn redwood, Ginkgo, and other trees that don't even exist here anymore. The climate had more diverse forests because it was a lot milder, so we went to look for species that were pre-historically native and for which we had some kind of evidence that they grew there during the last intense period of global warming. Sometimes we couldn't get the exact species, but we would get a close relative.

There was thought given to what would be likely to survive in an unmaintained area. If you have a garden, you can weed and water, but if planting in an industrial clear-cut you're gonna have to plant things that are gonna make it, so really delicious leafy things might not if there's heavy deer browsing. We chose species that we thought would be robust enough to take being stuck in the ground and forgotten about.

British Columbia during the Eocene had a more tropical climate, which hasn't yet changed back to that state. How did the plants that were adapted to a more tropical climate fare in your current iteration of this project?
Nobody knows what the climate's gonna do, other than get hotter. On the West Coast, it seems to be headed toward drought and forest fires. In the Northeast, we're getting more hurricanes and rain—too much in many cases. Planting a wide variety of species is a strategy— if certain plants like wet summers they should do better, or plants more tolerant of drought and fire that like more heat or more winter moisture should do well. As it turned out, the climate of that area is getting hotter and drier in the summer and wetter in the winter, so species that like summer rain did not enjoy what was happening.

There were also record droughts after we planted the trees, so many survived, but they didn't grow much and more than a few got chomped down because they weren't being tended. There are more extremes now, there's more chaos in the system. In the Eocene, it was more subtropical in a sense—it would be a little bit like northern Florida or that kind

of humidity, with much more rain in the growing season. The trees back then were far more diverse and the winters didn't really get much frost by the looks of things, and they're still getting some frost now. So, we didn't really plant super tropical stuff, although some palm trees are hardy in that area.

We hedged our bets and went for a wide spectrum. Now, as it turned out, the trees that did the best hands down are the coast redwood because in California they're getting set back by forest fires and drought. In coastal British Columbia the climate's getting drier, and the coast redwood is more tolerant of summer drought and winter rain. They're also able to feed from fog.

The other species that did quite well was the black walnut, which is now native to the Northeast, but there were similar, prehistoric walnuts once found in the West. Walnuts seemed a good candidate for survival because they have a deep tap root. They can get way down. The sequoia did reasonably well too on certain sites: gravelly, not too much moisture. Those three were the winners and some of the other ones like the Gingko, which we had high hopes for, were not so good. Dawn redwoods, which are a living fossil rediscovered in China in the 1940s, were killed by beavers but did spectacularly well elsewhere on the island. Details matter—anybody who is a professional tree planter will tell you about microsites. It really depends on exactly where a tree is planted on the land.

Assisted migration, like you mentioned, can be perceived as the planned transportation of non-native species. How have you entered into that dialogue with this work?
When we started that work, it was much more controversial than it is now. In fact, there were

the people who held the covenant on the land that were very opposed to what we were doing—we had to cut a deal with them to only do it on 30 percent of the land and not the whole property, even though it was all deforested. But now there are many similar projects happening here in the United States, like in New York where they are planting large test plots of trees from just a bit further south. Folks who are responsible for urban trees, like street trees, are no longer choosing historically native trees: in Massachusetts, the iconic New England tree is the beautiful sugar maple, but they're not planting those much anymore because they suffer with the increasing summer temperatures that exacerbate the urban heat island effect. They're planting trees from the Appalachians, Central South, the Mid-Atlantic, and further south—things like sweetgum and yellowwood—that are not recently native here, but have evolved to survive the hotter summers. Native forest lands are not, particularly when they're in stressful situations like urban landscapes. I would argue industrial clear-cuts are also very stressful situations. And where there is stress, pests are attracted and take advantage of the tree's reduced resistance.

So now there's this serious discussion: Should we move eastern hemlocks further north and bring in more southerly trees? This is being done in various places at scale but the details matter. If you were to bring a tree from Kentucky to Massachusetts, or from California to British Columbia, that would be sort of reasonable, but if you start to bring in something completely alien, that has never even prehistorically been in the landscape… There was a big issue in San Francisco with eucalyptus trees, for example, because they have never grown naturally in California or in the northern hemisphere. They're not in the fossil record, they're not native, and turn out to be somewhat invasive. That's where you have to be careful.

And so saying, bring in all plants from anywhere is not necessarily a good idea. But if things were either prehistorically native or were native in adjacent areas that are a little bit warmer, then I think it's fair game. But this is where judgment and consideration of what you're bringing in and where it grows now, comes in. And it is controversial. But my opinion is, we broke the climate so we need to deal with the consequences to the landscape.

I see this as range expansion, and trees can adjust their range. I'm confident in the coast redwood, because I know it grew fifty-five million years ago in British Columbia. The ecosystem where they are found natively historically is not that different. If you drive down from the Canadian border to California where the redwoods start, just south of the Oregon border, you'll see species that follow the whole range. Douglas fir can be found along with redwood and red cedar, depending on where you go. There are microcommunities that shift in composition as you move north and south.

I think the best strategy is to look at what's growing close by and see if it will grow a little further north, because the thing with the climate is that it too is an artifact. Here we are bringing a tree from California to Canada and planting it. That seems artificial, but climate change is artificial. We broke the climate. Human beings did that. It's not like it just happened on its own. We were the ones who pumped carbon dioxide in the atmosphere through excessive consumption of fossil fuel, although, it has to be said affluent parts of the world have a much higher per capita carbon footprint. I believe that we do have a responsi-

Oliver Kellhammer *Cunninghamia on Cortes Island, British Columbia*

bility to try out these assisted migration experiments to see if they work in confined areas—everywhere, but just to see if we can create some beautiful redwood forests in Canada, to replace the ones that are dying out in California. It's possible that the southern part of the range will no longer be viable for some of these trees because of the increased heat and frequency of fires and that's tragic and horrible. But on the other hand, if there are tons of redwoods growing north of the border—maybe that's something, and the species won't go extinct.

What can you share about the mycorrhizal connections that form underground between these trees as they settle into the land?
We're very lucky, as the owners of the land are Rupert Sheldrake, and his son Merlin, who wrote that amazing book *Entangled Life*. He is a soil mycorrhizal bacteria specialist who is taking samples and will be doing an ongoing study of this. Implementing my concept was possible only with their generosity and collaboration.

The temperate rainforest is a mycorrhizal-dependent forest. The soils are actually not very fertile. If you dig a hole, you'll see glaciated soils; the rain washes out a lot of the nutrients. However, you have these giant trees—how does that work? It's because of the mycorrhizal associates that are actually extracting nutrients and sharing nutrients in the forest.

One of the practices still happening in the Pacific Northwest is industrial scale clear-cutting, and then the slash and topsoil are burned in order to get a quick potassium hit from the ashes, basically incinerating most of the biomass and the all-important mycorrhizal life. That's like a sugar high: you're feeling tired and eat a chocolate bar and that's good for like,

twenty minutes, and then you basically pass out. That's what the forest industry continues to do—and they put bags of chemical fertilizer next to the seedlings they plant which have caused all kinds of health problems for tree planters. So my hope is that the redwoods we planted will have the memory to deal with the existing mycorrhizal life in that soil. So far, it seems, they are quite at home. They seem to be really benefiting from what's already there. And because coast redwoods and sequoias do grow in association with the leftover native plants still on the clear-cut, like the cedar and the Douglas fir, they may be getting inoculated by them—and so they should do just fine.

Next door is a property owned by the famous mycologist Paul Stamets. Paul has been experimenting with introducing beneficial mycelium in mulch form around seedling trees on the land. This is very active mycelium inoculation work but our approach is more passive. I'm really curious to see the specifics of Merlin's research. He has the scientific knowledge to actually find out what the mycelium are doing and what species they are.

This really is a new area in terms of understanding climate change because mycelium is, in many cases, a carbon absorber. So a lot of the carbon absorption of these forests is not just by the trees, but by the ground and it's the bacteria and the mycelium that really do bioaccumulate a lot of carbon and keep it from warming the climate. We're hoping that we can do more research over time, and this is gonna be an ongoing thing where Merlin will take samples and put them in the lab and actually find out what's going on there.

It's invisible to most of us. Finding out what specific species are there and how they've evolved or adapted throughout these different epochs is expansive to think about.
The interesting thing about it is that mushrooms have survived climate change before. There were periods of great extinction where there was volcanic eruption or asteroid impacts. There was global darkness. There was no sun and all the plants died, but that's when the mycelium provided—they could metabolize all the dead plants, all the charred stumps. They could metabolize that and actually create food for stuff that was still around. They are in a way survivors and healers. They can create a scaffold of fertility for things to regenerate after these horrible events come and go.

You come to this work as an artist and a permaculturalist and started this as a sort of bold conceptual gesture. How have your motivations or sensibilities around the project shifted and changed over time?
I never saw it as a particularly bold gesture, but I was surprised at how other people perceived it as kind of controversial. It seemed to me a pretty simple idea, but it is controversial. And now I think the world is kind of catching up to these ideas. I mean, often artists do things because we are free to be imaginative and unaccountable to institutions in the same kind of way as others. It's like, I just approached Rupert and said, "Hey, let's just do this and see what works." And he happens to be a well-known scientist, and his son a famous mycologist, botanist, and writer. So we were able to have the freedom to just go ahead, to do it.

Once Rupert agreed, the concept was just a question of raising enough money to buy a whole lot of trees and get' em in the ground, basically. So the freedom of being an artist is

good, but there's a whole tradition of this kind of land artwork. When I first went to New York in 1980, there was a very lovely work called *Time Landscape* at LaGuardia Place and West Hudson by an artist called Alan Sonfist who had planted on the site native vegetation that would have grown in Manhattan before European settler colonialism. It was very bizarre back then, these abject little shrubs and a vacant lot full of garbage. But I was really moved by it. And now, of course, all these years later, it is a beautiful park with giant trees.

Sonfist planted stuff that the built environment had already obliterated. So it is regarded as a public artwork, but it also functions as a kind of contemplative space where you can meditate on how things change over time. That's why it's called *Time Landscape*, because you're thinking about time as much as you are plants. So I was always really moved by how plants have this ability to bring us together and give us a sense of stewardship, but they do so much else for us too. I mean, we take care of them, they take care of us, right? This idea of a symbiotic relationship between ourselves and plants—a lot of artists have worked with that over the years.

Joseph Beuys was a big influence on me as well. He did this piece called *7000 Oaks* where instead of spending a lot of money on a monument he planted oak trees all through the German city of Kassel with big rocks beside them so they couldn't be taken down. His idea was that these trees are a legacy, that through reforesting a city and bringing back nature, we recreate the commons, as opposed to enclosing more property. Our redwood and walnut forest is not being planted to be cut down like the one that was clear-cut before it. It's gonna be a kind of museum, in a way, hopefully, for people hundreds of years from now who might come upon it and go: "Oh my God, what the hell? These giant redwood trees… What are they doing here? How are they growing so well?"

Hopefully people will think about how, back in the day, an artist and his scientist friends did this experiment when climate change was really starting to accelerate—a positive gesture that there might be a future. Climate change and global warming are catastrophic, but I think it's on us to reimagine how we're gonna deal with it, as opposed to just freaking out and hiding under the bed. ○

Vincent Waring *Young Redwood*

Rose Robinson

ON THE BRINK OF WINTER

In hours, we'll be starved of sun,
and tomorrow, even earlier so.
Our fingers will be colder,
and our toes will grow numb a little quicker.
We'll wake to snowflakes
and sleep to the creaks of the wood stove,
but for now we fill our pockets
with unborn daffodils,
our baskets with garlic cloves,
tuck them into the earth while it's still light.
Sleep will feel so sweet, as will the waking;
hunkered down and waiting to be
ready to lift our heads.

October 17, 2020
Two Harbors, Minnesota

ROUTES/ROOTS

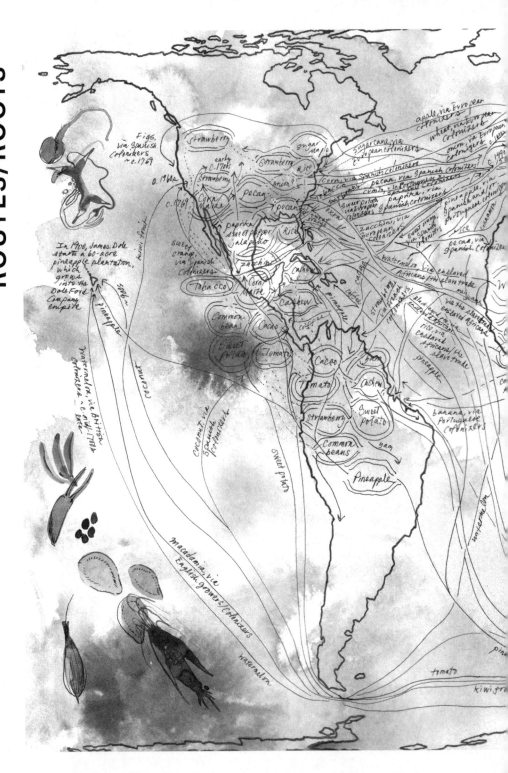

Tony VanWinkle

LIVING WITH AMBIGUITY AND *AKEBIA*

Invasive Species and Trickster Ecologies

Amid the suburban sprawl of western Guilford County, North Carolina, Guilford College Woods is a 240-acre preserve of mixed Piedmont forest in a sea of commercial and residential development. The Guilford Woods is simultaneously a refuge for local biodiversity and a major host site for invasive species. Among the most pervasive of these is the East Asian vining plant, *Akebia quinata*, commonly known as chocolate vine for its fragrant, early spring flowers.

The plant was introduced to North America in 1845. Charles S. Sargent, director of Harvard University's Arnold Arboretum and editor of the journal Garden and Forest, wrote in 1891 that A. quinata was by that time a common plant, "admired for its abundant dark green, digitate leaves . . . and for the abundant and curiously formed rosy purple flowers." The entry describes the plant's relatively underappreciated fruit and mentions its culinary usage in Japan, but characterizes it as "insipid" and unpalatable, concluding that "akebia fruit will probably never be valued in this country except as an ornamental."[1]

The 1925 catalog of the Greensboro-based Lindley Nursery Company lists the plant under the "Deciduous Climbing Vines" category. The catalog entry calls the plant a "popular

Japanese climber with beautiful foliage, almost evergreen," noting that its flowers are "peculiarly shaped" and are produced in April.[2] From its introduction into urban and suburban gardens, the plant came to populate local woodlands by its own botanical agency. Today, it is among the most aggressive invasive species in North Carolina and the greater mid-Atlantic, smothering trees, engulfing ground-level and understory habitats in a dense and impenetrable tangle of vine and leaf.

When discussing common control strategies in its contemporary literature on "alien" plants, the Plant Conservation Alliance devotes the bulk of its attention to several possible chemical controls.[3] These include various applications of broad-spectrum herbicides, including triclopyr and glyphosate. Such strategies are common in the invasive management world, but raise sticky ethical and moral questions for some of us. Which is doing greater damage in the world—fiveleaf akebia or Monsanto? Is the use of the poisons produced by the latter to eradicate the former an act of love or violence?

David Pellow states that pesticides "are produced for the expressed purpose of killing something, of ending the lives of plants or insects deemed to be invasive or alien."[4] It can

University of Alabama Herbarium   Akebia quinata *specimen* →

be argued that the logic of pesticide use, whether in agricultural or silvicultural contexts, is about practicing control. By employing the pesticide industry's tools, one is invariably deploying its violent infrastructures. That such infrastructures can be used for ecological simplification (as in the plantation model) or ecosystem restoration (as in the eradication of invasive species) is a rather difficult ontological problem.[5] To borrow from Max Liboiron's harm-violence continuum, the system that produces both the problem—invasive species introduced through botanical commerce and colonialist enterprise—and the purported solution—herbicidal poisons—begets continued systemic violence.[6] Akebia creates harms somewhat equivalent to the ubiquity of toxins that harm human bodies.

Lardizabalaceae
FLORA OF ALABAMA

Akebia quinata (Houtt.) Decne.

JEFFERSON COUNTY. Covering roadside and sloped bank like Kudzu at the junction of US-280 and Green Valley Road in Cahaba Heights. Birmingham-Big Canoe Valley district of the Ridge & Valley province. Vernacular name: Chocolate-Vine.
33° 27' 43"N, 86° 45' 22"W Birmingham South Quad

Perhaps a way out of the competing moral certitudes that such dilemmas generate can be found in what we might think of as trickster ecologies. This perspective applies our present socio-ecological quandaries to the teachings embedded in traditional trickster stories. For our present realities are characterized first and foremost by constant change, contingency, and ambiguity—precisely the domains where trickster consciousness thrives. In various Native American traditions, these stories might revolve around Coyote, Raven, or Rabbit.

These figures are often cunning shape-shifters, displaying reckless, vulgar, sometimes foolish, yet generative and adaptive behavior. Trickster stories help us reckon with the uncertainties that challenge rigid human categories in a constantly changing world. Scholars Thomas and Patricia Thornton assert that trickster "remind[s] us of the futility of a managerialism that governs only for control and stability without proper consideration of relational feedbacks and the dynamic and anarchic forces in nature."[7] As Ursula K. Le Guin characterized the function of Coyote in her own work as a writer of science fiction, "Coyote is an anarchist. She can confuse all civilized ideas simply by trotting through. And she always fools the pompous. Just when your ideas begin to get all nicely arranged and squared off, she messes them up. Things are never going to be neat, that's one thing you can count on."[8]

Rarámuri (Tarahumara) scholar and writer Enrique Salmón explains, "For Native peoples, when we have new plants—invasive species—show up, we don't think of them in a negative way. We think of them as part of the natural process, and then we come up with ways to incorporate them and adapt to them in our practices and our stories."[9] This ethic of accommodation, Salmón reminds us, reflects

291

University of Alabama Herbarium Akebia quinata *specimen*

millennia of Indigenous biocultural memory that has borne witness to unending cycles of change and reconstitution of the biotic communities of Turtle Island/Abya Yala. This—in contrast to settler colonial fictions of "pristine" continental landscapes at the time of contact, whose authenticity can and/or must

be restored—can serve as an alternative temporal benchmark in our reckonings with the scale of change.

When in bloom, the fragrance of akebia is delicious. Its fruit provides an edible gift, even if one outside of the normative gastronomic repertoire of contemporary North America. It reminds me of the many times I've reflected on the place of non-native plants that have become deeply interdigitated in the sensory ecologies of place. The intoxicating fragrance of invasive honeysuckle vine, for example, has been a part of the olfactory ecology of North Carolina and the greater Southeast for multiple generations now. That summer evening perfume is as much a part of that world as anything else.

Which all raises the question of how, or if, we might come to love species like akebia. If not love them, then at least tolerate them. Perhaps even ask what kinds of emergent trickster ecologies they might shepherd into existence and what kinds of productive nonviolent relationships we might forge with these newest occupants of our moral and natural landscapes. As crises and shared existential uncertainties mount, these emergent entanglements and biological formations encapsulate a parallel to scholar Michele Murphy's concept of Alterlife: "Life damaged, life persistent, life otherwise."[10] Andean peoples recognize this time we are living through as a Pachakuti, a transition between two ages. This is nothing short of a cosmological birthing event: miraculous and painful all at once, full of ambiguity and possibility. Perhaps it is time for a critical revitalization of trickster wisdom to guide us through these terrible and transcendent new becomings. ○

Notes

1. Charles S. Sargent, "Plant Notes: The Fruits of Akebia Quinata," *Garden and Forest* 161, vol. 4, (December/January 1891): 136.

2. J. Van. Lindley Nursery Company Catalog (1925).

3. Jil M. Swearingen, Adrienne Reese, and Robert E. Lyons, "Fact Sheet: Fiveleaf Akebia," Plant Conservation Alliance's Alien Plant Working Group, *invasive.org/alien/fact/pdf/akqu1.pdf* (2006).

4. David Naguib Pellow, *Resisting Global Toxics: Transnational Movements for Environmental Justice* (Cambridge: MIT Press, 2007): 149.

5. The plantation, as both a historical and contemporary model for arranging human-nature relations, radically simplifies complex natural and social systems. For example, the plantation replaces complex ecologies and social relations with monocultures and exploitative labor relations that serve the singular purpose of capital accumulation. For further theorization of this concept, see Donna Haraway, "Anthropocene, Capitalocene, Plantationocene, Chthulucene: Making Kin," *Environmental Humanities* 6 (2015).

6. Max Libioron, *Pollution is Colonialism* (Durham: Duke University Press, 2021).

7. Thomas F. Thornton and Patricia M. Thornton, "The Mutable, The Mythical, and the Managerial: Raven Narratives and the Anthropocene," *Environment and Society 6* (2015): 66–86.

8. Ursula K. Le Guin, *Buffalo Gals and Other Animal Presences* (New York: Roc Books, 1994).

9. Ayana Young, "Enrique Salmón on Moral Landscapes Amidst Changing Ecologies," March 20, 2021, in *For the Wild*, podcast, MP3 audio.

10. Michelle Murphy, "Against Population, Towards Alterlife," in Adele Clark and Donna Haraway, eds., *Making Kin Not Population: Reconceiving Generations* (Chicago: University of Chicago Press, 2018).

11. Lucia Stavig, "Experiencing Corona Virus in the Andes," in Rebecca Irons and Sahra Gibbon, eds., *Consciously Quarantined: A COVID-19 response from the Social Sciences*, May 2020. *medanthucl.com/2020/05/04/experiencing-corona-virus-in-the-andes*.

Ernest Henry Wilson Akebia lobata *Massachusetts, Milton* (image courtesy of Arnold Arboretum Horticultural Library) ↑

Ernest Henry Wilson Davidia involucrata Kew, *England* (image courtesy of Arnold Arboretum Horticultural Library) ↑

Alex Arzt

BECOMING NATURALIZED TO PLACE[1]

California's Feral Cabbages

The feral cabbages (*Brassica oleracea var. oleracea*) of California's coast are a food crop resilient to extreme weather, toxic air, and contaminated soil. They often grow on the edges and cracks of lighthouses and littoral military bases, and, after more than a hundred years growing without intervention, have become naturalized on the coast.[2] Since 2018, I have been growing them in the yard in my rented house in East Oakland, where they have thrived in soil that is both highly fertile and contaminated with lead. They readily embrace the conditions of the garden as well as the rugged seaside cliffs, and my work with these plants has opened up a deepened awareness of place and human-plant relationships.

First, a few helpful definitions. A **wild** plant has not been artificially selected by humans, so its genes have not been intentionally manipulated. **Domesticated** plants are "human-propagated organisms [that] adapt to humans and the environments they provide."[3] A **cultivar** is "an organism and especially one of an agricultural or horticultural variety or strain originating and persistent under cultivation."[4] **Feral** plants descend from artificially selected ancestors but live outside of human cultivation.[5] With *Brassica oleracea*, it can be difficult to parse out the wild from the feral or pin them down into any single category. A perennial here, the cabbages have continued to flourish in a sometimes chaotic neighborhood where we face prolonged drought, heat waves, and wildfire smoke capable of blocking out the sun. These are the realities of living in urban

California, and it is urgent to find food crops that can withstand such bodily stressors.

Farmers and amateur gardeners can help discover or create these plants, not just university-funded plant scientists. After all, most of the vegetables we eat today came from the sustained attention of early agricultural peoples, who selected seed from one plant over another looking for sweeter fruit, bigger leaves, or higher yields and repeated this act for thousands of years.[6] My long-term engagement with feral cabbages offers a knowledge of the land where they grow and where I live through a process of making what William Least Heat-Moon termed "a deep map," or "a way to be conscious of a place in such a manner as to hold multiple layers of understanding."[7] The cabbages have cemented a personal place-based land ethic: giving sustained attention to and caring for plants can reveal unknown histories and relationships, novel ways of being, and new avenues of survival.

There are at least five morphologically unique feral cabbage populations on the California coast, four of which I am cultivating: from Point Bonita Lighthouse, Point Montara Lighthouse, Pigeon Point Lighthouse, Mendocino Headlands, and Half Moon Bay State Beach.[8] Using the methods in Carol Deppe's *Breed Your Own Vegetable Varieties*, I am breeding the five all together and selecting for the plants that do the best under the most challenging conditions with the intention of sharing the seeds with other gardeners and farmers.[9] In spring 2021, I created a cross between Point Bonita and Mendocino. This second season will incorporate Point Montara into the mix, and next year will be Half Moon Bay and Pigeon Point.

The Point Bonita cabbage most likely escaped the garden of the wife of a lighthouse keeper at the turn of the century and has since established an isolated, naturalized cabbage community on a narrow basalt headland overlooking the Pacific Ocean and the San Francisco Bay.[10] This feral cabbage's blanket taxonomic classification is *Brassica oleracea var. oleracea*, though the plants themselves might vary genetically and morphologically depending on where they grow and from which cultivars they originated.[11]

Their temporal and genetic history is inextricably linked to humans in a co-evolutionary relationship. Research suggests the cultivation of wild cabbage began in western Syria between 3250–2970 BP and has since been selectively bred into broccoli, cabbage, cauliflower, kohlrabi and brussels sprouts, to name a few.[12] They escape cultivation and naturalize themselves in coastal environments similar to their origins and can just as easily revert back to their secondary habitats: cultivated gardens.[13] Maybe the plants possess a kind of genetic memory of three thousand years of human hands harvesting their leaves, watering them and protecting them from harm, feeding their soil, pollinating them, spreading their seeds, and ensuring their survival.

In 2017, I moved into the house where I live now. I cleaned up the yard and put in plants from friends, gardening jobs, and seed packets. My early philosophy—stick it in the ground and see what happens—was aided by the moderate climate of the Bay Area and the inherent fertility of the yard's soil. Before I moved in, tenants largely neglected the outdoor space, and so the nasturtiums, oxalis, fig tree, and termite-infested peppercorn tree maintained themselves.[14] My sister, who lived here before me, added a plum tree, cedar, bird-of-paradise, and a dye garden.

I did not eat the cabbages, caring for them as one might a pet or a rose bush. Witnessing the Point Bonita cabbages grow huge and healthy away from their oceanside perches initiated a journey to find their origin. It became apparent how difficult it is to read a landscape where there is little obvious trace of what came before, which ecologists call "shifting baseline syndrome."[15] Some plants persist when the people who stewarded them leave and are forgotten. The same phenomena applies to this lot in Oakland—and to all land with a known or unknown history of human cultivation.

According to an 1852 map, this lot once existed between a salt marsh on the tidal banks of San Antonio Creek and **encinal**, evergreen oak woodlands.[16, 17] The Ohlone inhabited this place in the territory of Huichin as their ancestral home for up to fourteen thousand years.[18] Around the time wild cabbages were first domesticated in Syria, the Ohlone were stewarding the fertile and abundant land now known as the East Bay by harvesting acorns, practicing controlled burns, hunting, foraging, and fishing.[19] Spanish colonists stole their land and displaced them to the Catholic missions beginning in the 1770s.[20, 21, 22] Beginning in the 1820s, it was granted to the Peralta family by the Spanish king and became part of a 44,800-acre cattle ranch.[23] The land changed hands throughout the 1800s, towns were established, and it finally became part of Oakland in 1872.

Built around 1878 in a neighborhood near the water, the house is an Italianate style one-story cottage that was originally painted with white lead paint.[24] It was foreclosed on

Alex Arzt *Front yard, April 2022*

once in 1880 and again in 2011.[25, 26] The original bay waterfronts with their ecologically diverse tidal marshes were filled in to make way for train tracks and then the freeway, where gasoline-powered cars spewed lead into the air until 1996.[27, 28] The neglect of the property contributed to both the toxicity and the fertility of the soil. Mark Stumpf, who owned the house between 1989 and 2011, told me he removed layers of garbage from the yard and under the house.[29] A gardener, builder, and painter, Mark planted a fig, pear, and an orange tree; had a small vegetable garden; and buried three of his Irish wolfhounds in the front yard, beneath where the cabbages grow today.

After moving in, I turned the yard into a semi-wild garden with cacti, ornamental flowers, and native plants, letting them manage the space themselves with minimal intervention, all with a preference for the cabbages. The soil itself is a bit sandy, easy to crumble in your hands, and medium dark brown like a light roast coffee. I didn't think to have the soil tested until June 2021, and the levels of minerals, elements, and metals listed in parts per million (ppm) were clues to the soil's history and memory—another layer towards forming a deep map.

It proved highly fertile and also off the charts for lead—around 800 ppm—therefore "not suitable for gardening."[30] Until then, I was largely ignorant of this ubiquitous urban problem and the resulting health implications, which include brain damage, digestive problems, and infertility.[31] These effects are most severe and lasting in children, whose bodies retain more lead than adults.[32] Learning of the contamination created a shift in my relationship to the land here, which is both poisonous and life-giving. I became afraid of it, and I wanted to immerse myself in it, to continue our

ongoing collaboration. In this dichotomy lies a confusing kind of grief.

The lead is in my blood and the cabbage's tissues. My blood lead level is 2 mcg/dL, slightly higher than the average adult level in 2016 of 0.92 mcg/dL.[33] The sampled cabbage leaves were in the range of 3 ppm. They were uptaking the heavy metal, and there is no standard safe lead level in food. My first instinct to let the plants be and not eat them regularly was a good one.

Despite the temporary nature of my stewardship and the fact that the soil will all likely be covered over eventually, the garden is healing the soil. The plants uptake lead, add organic matter, and create biological diversity. Organic matter reduces the amount of lead available to the plants because of the way it chemically binds to the metal.[34] There is more research to be done. Since 2020, I've harvested two quart jars full of seeds each season, and I'm currently dispersing them through seed packets and to whoever asks for them. The house finches perch on the stalks and eat the seeds, hopefully spreading them further afield. I once saw a volunteer Red Russian kale *(Brassica napus var. pabularia)* growing in the sidewalk around the corner, further evidence of the Brassica family's adaptability.[35]

While spending time away in Oregon recently, I listened to the audiobook of Robin Wall Kimmerer's *Braiding Sweetgrass* on long walks not far from a salt marsh. A plant scientist, professor, and member of the Citizen Potawatomi Nation, she recounts sitting in her garden one day, where she had an epiphany that the land loved her back through "material and spiritual" relationships of reciprocity.[36] Her realization ran through me. "Food plants and people act as selective forces on each other's evolution—the thriving of one in the best interest of the other," she writes. "This, to me, sounds a bit like love."[37] The thought of the cabbage garden in Oakland loving me back brought me to tears of overwhelm and gratitude under a big desert sky on an ancient lakebed.

With love, of course, comes inevitable loss and grief, and for gardeners who rent, this creates a conundrum. In a region with some of the most expensive housing in the country, the value of this lot, which has risen 30 percent over the last decade, has nothing to do with soil fertility—but value is subjective and relational.[38] I will eventually move away and say goodbye to the garden. With this knowledge, I diligently save the cabbage and flower seeds in big plastic bins to thresh each fall. I have a mental map of where bulbs are planted so I can dig them up and bring them to the next place. The first Point Bonita cabbages from 2018 lived their full perennial lifespan of two-and-a-half years and are decomposing into the soil. One of its clones and three of its progeny will continue as the second generation of Oakland feral cabbages. Even after I leave, I hope their seeds continue to germinate after the winter rains.

Since my time here is temporary, I'm preparing to take the gifts of this place with me. But it has rooted in my body and consciousness. The smell of hydrogen sulfide from last season's decomposing cabbage leaves now reminds me of home. According to Least Heat-Moon, "to know a place in any real and lasting way is sooner or later to dream it. That's how we come to belong to it in the deepest sense."[39] I have recurring dreams of the landlord mowing down the garden and destroying the cabbages. Last time, he replaced them with a new condo-style landscape design, and in the dream I cried on my knees like a grieving mother. The plants and I belong to this place.

Last season, in an attempt to reduce the inevitable aphid infestations that come with repeated brassica plantings, I set out to create a symbiotic cabbage garden in Oakland by planting phacelia, buckwheat, and alyssum, which I learned attract predatory and parasitoid wasps.[40] The wasps came and I noticed an increase of aphid mummies under the cabbage leaves. Later on a walk to Point Bonita, I noticed that wild buckwheat, phacelia, and alyssum were already growing around the feral cabbages, creating cooperative ecological relationships among other native and non-native species. They were naturalizing to place in the way that Kimmerer proposes "[throws] off the mind-set of the immigrants."[41] The feral cabbages offer a kind of land ethic as immigrants themselves by establishing communities with the common goals of mutuality, reciprocity, and cooperation, as opposed to competing, poisoning, and smothering a place's indigenous flora—the characteristics of what we call "invasive" or "noxious" plants. Like the common plantain that Kimmerer describes, the cabbages have become naturalized to California.

The garden in Oakland attests to the resilience of the cabbages and their allies who have created a web of life that is beyond my observable powers via birds, insects, fungi, rodents, and soil microbes. Feral cabbages reveal a path towards survival through discovering their ecological niches, forming cooperative communities, and creating multispecies partnerships, including with me and other humans as dispersal agents and caretakers. They are worthy of our attention and study as a means to connect us further to the places where we live and grow and to elucidate an ancient and ongoing human-plant relationship.

My research and process towards building this deep map continues. Looking ahead, I am seeking space to grow the plants in a dedicated,

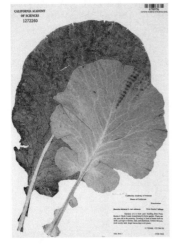

Alex Arzt *Point Bonita cabbage in Oakland, California*

uncontaminated space to continue the feral hybridization project. They are already creating unique generations of hybrids further adapted to the urban environment. As a potential source of resilient, nutritious food and as connectors to our plant communities, the feral cabbages can both feed us and teach us. ○

California Academy of Sciences *Brassica olerecea specimen* ↖ ↑

1. Robin Wall Kimmerer, *Braiding Sweet-grass* (Minneapolis: Milkweed Editions, 2013).

2. Missouri Botanical Garden, "*Brassica oleracea L. var. oleracea.*," *Tropicos.org*, tropicos.org/name/4105289.

3. Eben Gering, Darren Incorvaia, Rie Henriksen, Jeffrey Conner, Thomas Getty, and Dominic Wright, "Getting Back to Nature: Fertilization in Animals and Plants," *Trends in Ecology and Evolution* 34, no. 12 (December 2019): 1138.

4. Merriam-Webster.com Dictionary, s.v. "cultivar".

5. Gering et al, 1138.

6. Noel Kingsbury, Hybrid: *The History and Science of Plant Breeding* (Chicago: The University of Chicago Press, 2009), 16–20.

7. Brett Bloom, "Deep Maps" in kollektiv orangotango+, ed., *This is Not an Atlas: A Global Collection of Counter-Cartographies* (Bielefeld: transcript Verlag, 2019), 300–306.

8. Alex Arzt, ed., "Wild Cabbages of the World," *iNaturalist*. inaturalist.org/projects/wild-cabbages-of-the-world.

9. Carol Deppe, *Breed Your Own Vegetable Varieties: The Gardener's and Farmer's Guide to Plant Breeding and Seed Saving* (Boston: Little, Brown, and Co., 1993), 311–314.

10. Norma Engel, *Three Beams of Light: Chronicles of a Lighthouse Keeper's Family* (San Diego: Tecolote Publications, 2004). 32–34.

11. Alex Arzt, "*Brassica oleracea L.*," *Global Biodiversity Information Facility*, gbif.org/occurrence/3490212343.

12. Makenzie E. Mabry, Sarah D. Turner-Hissong, Evan Y. Gallagher, Alex C. McAlvay, Hong An, Patrick P. Edger, Jonathan D. Moore, et al., "The Evolutionary History of Wild, Domesticated, and Feral Brassica Oleracea (Brassicaceae)," *Molecular Biology and Evolution* 38, no. 10 (June 2021): 4419–4434.

13. John Thomas Howell, *Marin Flora: Manual of the Flowering Plants and Ferns of Marin County, California* (Berkeley: University of California Press, 1970), 145.

14. Mark Stumpf, conversation with the author, January 17, 2022.

15. Anna Lowenhaupt Tsing, Heather Anne Swanson, Elaine Gan, and Nils Bubandt, eds., *Arts of Living on a Damaged Planet: Ghosts of the Anthropocene* (Minneapolis: University of Minnesota Press, 2017), G6.

16. US Forest Service, "Evergreen Oak Woodlands or Encinal," *fs.fed.us/wildflowers/beauty/Sky_Islands/communities/evergreenoaks.html*.

17. Allexey Waldemar Von Scmidt, "Plan of Rancho de-San-Antonio," 1852, UC Berkeley Bancroft Library, *oac.cdlib.org/ark:/13030/hb5f59n96n/?brand=oac4*.

18. Randal Milliken, Lawrence H. Shoup, and Beverly R. Ortiz, *Ohlone/Costanoan Indians of the San Francisco Peninsula and their Neighbors, Yesterday and Today*, (Oakland: Archeological and History Consultants, 2009), 71.

19. Malcolm Margolin, *The Ohlone Way: Indian Life in the San Francisco-Monterey Bay Area* (Berkeley: Heyday Books, 2014), 23–63.

20. Milliken et al, 70, 87, 88.

21. UC Berkeley Centers for Educational Justice & Community Engagement, "Ohlone Land," 2022.

22. Margolin, 59.

23. Friends of Peralta Hacienda Historical Park, "Chronology of Events," *peraltahacienda.org/pages/main.php?page-id=71&pagecategory=3*.

24. Oakland History Center, Property Records, 1878.

25. "Sheriff's Sale," *Oakland Tribune*, April 22, 1880.

26. Mark Stumpf, conversation with the author, January 17, 2022.

27. Janet M. Sowers, "San Antonio Creek Watershed Map," in *Creek & Watershed Map of Oakland & Berkeley* (Oakland: The Oakland Museum of California, 1993), *explore.museumca.org/creeks/1180-OMSAntonio.html*.

28. Jesse Stolark, "Fact Sheet: A Brief History of Octane in Gasoline: From Lead to Ethanol," Environmental and Energy Study Institute, March 30, 2016.

29. Mark Stumpf, conversation with the author, January 17, 2022.

30. Wallace Labs, Soil Test, June 8, 2021.

31. Christian Warren, *Brush with Death: A Social History of Lead Poisoning* (Baltimore: Johns Hopkins University Press, 2001), 16.

32. "Health Effects of Lead Exposure," Centers for Disease Control and Prevention, March 9, 2022.

33. "ABLES - Reference Blood Lead Levels (BLLS) for Adults in the US," Centers for Disease Control and Prevention (The National Institute for Occupational Safety and Health (NIOSH), February 23, 2021.

34. Chammi P. Attanayake, Ganga M. Hettiarachchi, Ashley Harms, DeAnn Presley, Sabine Martin, and Gary M. Pierzynski, "Field Evaluations on Soil Plant Transfer of Lead from an Urban Garden Soil," *Journal of Environmental Quality* 43, no. 2, June 2014: 475–487.

35. Alex Arzt, "Rapeseed (Brassica napus)," *iNaturalist*, inaturalist.org/observations/42614838.

36. Kimmerer, 123.

37. Kimmerer, 124.

38. Zillow, 'Zestimate' and Sales Records.

39. William Least Heat-Moon, *PrairyErth: A Deep Map* (Boston: Houghton Mifflin, 1991), 105.

40. Michele Meder, Genevieve Higgins, and Susan B. Scheufele, "Attracting Beneficial Insects to Reduce Cabbage Aphid Population Size," (Brassica Pest Collaborative, 2018).

41. Kimmerer, 215.

Dimitra Skandali   *Oceans and Seas II, V. 4*
Oceans and Seas are large scale sculptures made from crocheted sea grass *(zostera marina)*,
found and given fishing nets, and found plastic strings.

Nanci Amaka

FETCHING WATER

How to stand still in living water:
Plant yourself.
Spread your toes. Press them into the soil or sand
Dance your feet into the earth
Like the Ede plant, you need to plant yourself in mud,
Take root and hold on.
Only when your feet are rooted, can you let go of your body
Turn your joints to water,
Surrender to the ebb and flow,
Remember you are also water.

Then

For as long as I could remember, we awoke before the sun to walk ancient paths around my village. Pathways softened by millennia of rainy seasons and packed down by humans and animals lured by this pied piper of sorts. Fresh water. We would wake and hike down to the valleys to find her awake and winking in the dawn light.

Along the path, a steady stream of human bodies wet from sweat and slosh; clutched and balanced jerrycans, clay pots, and buckets. We met and greeted each other. "*Unu awunno!*" May you not die. Live! Along the path news was shared. Our social selves, entangled; our memories, woven together.

At the bottom of the valleys were the rivers, life. Everyone entered the water at different points to commune with her in various ways: to bathe, launder, drink, gossip, play, and fish. We would then fill our containers and begin the arduous trek back to the villages built high above the flood lines. With our vessels full of water, we'd trudge, careful-footed, pulling ourselves and the water uphill in a determination for life. Life for the elders whose joints have been ground by the journey before us. Life for the garden. Life for the chickens and goats. For cooking our meals, for washing the rest of the day. For drinking and filling ourselves when the sun is high and awake.

Umunnoho is a tropical rainforest village in southeastern Nigeria. A place older than memory where I spent my formative years. My parents were born at the formation of the post-colonized nation of Nigeria, in a generation of civil war survivors. A generation who grew up in one world that was quickly swallowed by a bigger, greedier one. Youth whose parents' lifestyles, they believed, lacked the glamor of globalized desires: a Western education, medicine, processed foods, ready-made clothes, indoor plumbing, and—most of all—money to buy it all. When what you want can only be bought, you need money and ways to earn it. So we fled, leaving the villages for education, to find jobs, and to earn money. A pied piper of sorts—this dream of a better world, a better way.

How to survive standing in living water:
Come back to your body.
Remember that you are also water,
Water pulsing around an anchor of bones,
Enveloped in viscera.

Now

These days I live in Honolulu, on a mountain one mile high. Some mornings, when nostalgia aches, I wake before the sun to trek to the bottom of the hill and stare at the unforgiving concrete pavement of what was once Wai`alae Stream. If there is any water in it, it is the remnants of rain from the night before, waiting stoically for the sun to rise and to be metamorphosed. In those moments, I feel like an alien creature shocked to find itself in a not-so foreign land. Full of commodities and no easy access to free, fresh water.

These days I cover my mouth in company. To catch my breath, fetch the moisture leaking out of me. Stopping, protecting, isolating it. Back in my village, my cousins, who migrated to Europe decades ago, have built a fence around our ancestral compound. All the way from Amsterdam, they built walls around their birthrights. The land is now a commodity: something owned. The paved roads leading to the villages are lined with unfinished mansions rotting in the tropical air, seeping into them, expanding and contracting in the African heat. You can trace the water's path by following the mossy trails down the cement bricks of half-tumbled walls into the wild earth. Half-realized dreams of diasporic descendants hoping to one day be reunited with the bones of their ancestors who have since released all their water back into the untamable ground.

A few days ago, I spoke with two friends about the idea of wild water. One mentioned she only encounters wild water while camping. The other has her own water-catchment system and is off-the-grid. Aside from worrying about toxins in the roofing material of her home poisoning the water, she hasn't yet had to worry about whether or not she will have water. She lives in a system of abundant water. But as ideal as this is, it still relies on initial capital to make it a reality. Regardless, the three of us agree that this was the best way forward.

I remember the dash to set out pails and buckets when it rained in the village to catch water. Especially memorable was the sensation of watching each droplet bounce and create a ripple that cascaded with other ripples—a mesmerizing and hypnotic vision, even in memory. I remember the joy of filling all our basins and knowing the crops and animals, ourselves included, would be nurtured. Moreover, I remember the relief of not needing to trek to the river for a few days.

In Between

For a long time after leaving my village in Nigeria, I lived in major metropolises: Chicago, San Francisco, Los Angeles, and Boston among them. Now, living in Honolulu, I can almost tell the difference in water… almost.

When you are enshrined in nature, it is easy to think that there has been a mass-deception of people to believe they have to rely on capitalism to survive. The truth is, in the belly of this tower of Babel we are building—this edifice of concrete and steel—there is no choice. Those who subsist far away from nature have to get their resources piped to them. Unfortunately, the feeding tubes come with their palms up, demanding compensation; and

Nanci Amaka   Icho Ije | *Standing Water*

the exchange is more abstract than a trek downhill to a river and back up.

Even here in the beautiful Kingdom of Hawai'i,[1] where water falls wild and clean from the sky. On these lands that were nurtured for generations by Indigenous Hawaiians, where water catchment is legal and encouraged; I still pay for water to be piped a mile uphill. Because, although I live in paradise, I do not own a home and do not have the funds or permissions to obtain nor install a water-catchment system. I cannot afford to call any land

mine. Yet the whole world is my home and I feel immense responsibility for it. An anxious, suspended existence.

As a child, I fetched water for the elders in my village because they had paved the paths before us to get to the rivers and streams. The older I get, the more I think about this: What pathways have I paved so far in my life? The scarier question, have I even paved any at all; or worse, continued down a path I know leads to risks of poisoned water?

The United States Navy Red Hill Bulk Fuel

306

Storage Facility is a few miles from my home and is in the middle of a drinking water pollution crisis. Their leaking fuel containers are located less than one hundred feet above aquifers that supply more than 75 percent of the freshwater used on the entire island of Oahu.[2]

On the other side of the planet there's an echo and it is of the massive oil leaks and pollution that destroyed the rivers of the Niger Delta region of Nigeria by the Royal Dutch Shell oil company and the subsequent political execution of environmental activist Ken Saro Wiwa in 1995 by Sani Abacha—a kleptocrat dictator.[3]

A decade ago, I visited Nigeria with my siblings and while taking a walk, we found a live fish in a puddle in the middle of the road. Miles away from any river. For a while, it seemed like a riddle. It wasn't until much later that we witnessed convoys of lorries hauling loads of sand from the riverbanks to building sites across town. It seems even rivers that haven't been destroyed by massive oil leaks cannot escape the effects of modernization.

Future

These mornings, I collect dewdrops from my houseplants. I keep them in a tiny vial, and nobody—including myself—knows why I'm doing this. There is no deep meaningful reason that I can access in my psyche, except that it makes me happy to see these dew drops when I wake up. My impulse to collect them feels raw, wild.

I can't help but think that every time humans have tamed forces of nature, the lasso used to wrangle it entangles us further into a web of co-reliance that is ultimately uncontrollable. Nature as a commodity compounds already complicated relationships between humans and their unquenching thirst for power and greed, as we humans are also wild things that act for the betterment of ourselves over the global good.

The last time I was in my ancestral village, a neighbor came to ask my aunt and uncle to build a gate for them to walk through the compound so that they could continue to follow the old paths to the rivers. Asking us to release our stubborn nostalgic grip on everything we grew up loving: all the treks at dawn filled with laughter and straining, all the joy of communing with family and neighbors, the shared struggles, and the feeling of belonging to a boundless, ancient system. They came to ask us to open the cage we had built around it all. So that they—who don't just visit but live in the villages still, who walk those paths daily, who keep it paved for future generations—may live. Begging that their lives not be suffocated by the detritus of the unsustainably globalized world that caused us to flee in the first place. ○

Notes

1. Calling this land 'The Kingdom of Hawai'i' allows room in the imagination for many other realities to grow while also respecting the history of the place and the will of its people.
2. Christina Jedra, "How the Red Hill Fuel System Has Threatened Oahu's Drinking Water For Decades," *Honolulu Civil Beat*, December 12, 2021.
3. Amnesty International, "Investigate Shell for complicity in murder, rape and torture," November 28, 2017.

Rachel Edwards

STAR SEED SYLLABLES

Summer

below Seven Stars stretched out across the Open Sky
treasure bodies alight emerald fields
crimson tendrils, golden blossoms
glistening pearls murmuring—
Destiny is in our hands:

Spring

quiet ground, still seeds
listening, tending
the Mother within every child
whispering—
One sound moves the world:

Autumn

gathering kindness into a large serving bowl
spoon some onto every plate at the table
add a side dish of forgiveness
saving some for yourself—
Empty once more:

Winter

before it's too late
becomes early, becoming time
to rest on our knees
covered in dirt, in Darkness—
Return:

Mo Walrath   *Opening the Earth: woven coffins as offerings*
for the beings they hold, the land that will receive them,
and the possibilities of healing through human hands and culture

Taylor Hanigosky

THE THICKET DWELLERS

The thicket dwellers make their nest in the bramble. They chop rocks to remind them of their power. They wash wool to pray for softness. They untangle knots to connect.

Usually, they spend their winter afternoons stomping down dead-standing branches and loosening naked vines from oak seedlings.

But one day, a windstorm was swelling in the South. It picked up excess speed and all the littered plastic bags from the sides of the interstates, and headed toward the thicket with gusto.

In the open fields all around the thicket, the soil was taken up and away. A swirling cloud of dust consumed the work-ers, the worry-ers, and the wait-ers who lived out in the open spaces and scrambled their identities, causing a spell of confusion. But the thicket held strong, and the dirt stayed.

After the storm, residual plastic bags collected in drifts, some as tall as the poplars at the edge of the thicket. The work-ers (who used to be the worry-ers) anxiously went out to gather all the plastic bags into bigger plastic bags. The worry-ers (who used to be the wait-ers) paced back and forth, paralyzed by a deep indecision of whether to help clean up or to do nothing. The wait-ers (who used to be the work-ers) sat and watched the plastic drifts with a gnawing, unresolved feeling.

Inside the thicket, the chaos of the plastic drifts was not felt through the chaos of coiling vines and overlapping twigs. But the thicket dwellers knew deeply their power to alter their surroundings. They remembered their prayers

for softness. They felt connected to the tangle.

They crawled out of the thicket and into the open each morning to gather an armful each of the plastic bags. They spent their afternoons carefully and methodically spinning the plastic bags into rope. It took a very long time, but the thicket dwellers had plenty of that. The work-ers grew frustrated that the thicket dwellers would not help with the disposal efforts. The worry-ers feared for the thicket dwellers' sanity. The wait-ers sat and watched.

When the plastic rope grew very long, and many of the bags had been spun up, the thicket dwellers took to running lines of rope back and forth. They looped around a lonely tree and passed back to criss-cross over a patch of bare soil many times. They crawled up the sides of the tallest empty building and back down, letting out lines of their plastic rope, like the silk of a spider, everywhere they went. Up flag poles, down lampposts, over dumpsters, and across parking lots until everything was tangled up to the thing next to it. The ground lay beneath a lattice of colorful rope.

Time mostly passed this way. The work-ers, worry-ers, and wait-ers began calling the thicket dwellers the weave-ers.

One day, a wild and angry storm was growing in the South. It picked up speed and the littered plastic bottles along the sides of the interstates. When the winds reached the plastic thicket, nothing blew away. The dirt, the trees, and the buildings stayed. The plastic bottles all got caught in the plastic web. When the storm passed, they hung high and scattered and captured the sunlight coming from behind the storm clouds.

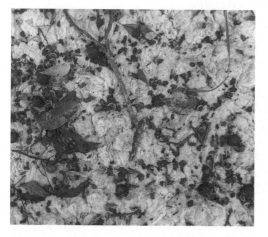

The bottles sparkled full of sun like dizzying mirrors and crystals. The dazzling light transfixed the work-ers, the worry-ers, and the wait-ers into a stupor. Birds from near and far caught wind of the dreamy thicket of colors and light and undertook a migratory pilgrimage en masse to be immersed in the wonder. With them, the warblers, the wrens, and the whippoorwills brought many seeds from many bearing fruits, and found many places to perch upon the criss-cross plastic rope. They planted their seeds and nutrient-dense droppings on the surface of the bare soil. Soon, the bare soil sprouted eager and hardy life that grew up through the plastic rope in a riot of green.

Some of the work-ers went around cleaning up the bird poop. Some of the worry-ers paced over and under the plastic rope lines, fretting about what would happen to the thicket if one of the lines snapped from the weight of the nesting birds. Some of the wait-ers sat amid the new and thick growth and pondered what they could no longer see. But some others buried a part of themselves in the loamy soil and crawled deeper still into the thicket, to find the original thicket dwellers.

They found them chopping rocks, washing wool, stomping down dead-standing branches, and loosening vines from around oak saplings. They joined together. ○

Acknowledging Violence

Hunger, work-life (im)balance,

culling, killing,

confronting ease

Morgan Whitehead

METABOLISM

Hunger, belly, moon.

Two white dogs in a cloud of feathers, pulpy mange
flecked across the dead and dying chickweed.

Tangled limbs of scrub oak
submit to each other, ashen with cold.

We find teeth in the woods but no skull.

The new moon is directionless, a seed,
and hurtles through the darkness, unseen.

Elizabeth Henderson

UNTO THYSELF BE TRUE

A Whole Life Approach to Resilience at Rock Steady Farm

In the United States, rural areas are not always welcoming of newcomers, especially if they are LGBTQIA or people of color. Defying deeply ingrained prejudices as well as the economic assumption that to pay the bills organic farms have to sell to high-end markets, Rock Steady Farm has been able to create a successful farm and welcoming community space with over half of its sales going to low-income households. By responding to the pandemic quickly and skillfully, the Rock Steady farmers have even been able to increase community support.

Late in 2015, Maggie Cheney and D Rooney established Rock Steady on twelve acres in Millerton, New York, a two-hour drive north of New York City. With a third partner, Angela DeFelice, who now operates as a financial advisor to the farm, they set out a complex social mission for an LGBTQIA-run cooperative, rooted in social justice. The partners proudly declare that they are "endlessly grateful to be who we are, and engaging in farming with both care for each other and the earth as best we can." Creating a space where they themselves are comfortable and providing that sense of openness and acceptance to others is central to their effort.

The Rock Steady approach to farm creation is the opposite of John Wayne-style individual-

ism. D and Maggie are methodical planners who take advantage of their own years of experience, the wisdom of farming elders, and counseling and advice available through their connections with the cooperative and social service communities. D and Maggie's initial marketing plan was to develop a community-supported agriculture (CSA) program with share payments on a sliding scale so that lower-income people could afford them, and to offset the lower prices of vegetables with sales of flowers that generate higher revenues. Rock Steady grew flowers for the first two years, but has put them on pause due to the rapid growth of the CSA propelled by the pandemic. A post on their Facebook page on April 7, 2021, declared, "CSA, we have actually sold out! Five hundred members strong, it's our largest CSA to date! Woah!"

To access the capital to start up Rock Steady, the partners were able to take out a one hundred thousand dollar loan from The Working World, a "non-extractive" financial lender. The farm has also received funding and business advice from Seed Commons, the Cooperative Development Initiative, Community Food Funders, and the 2020 Food Movement Support Fund The farm also worked with a dozen other grant-making organizations and private foundations

Rock Steady Farm *Rock Steady barn and rainbow*

in partnership with The Watershed Center—whose land they lease—acting as their fiscal sponsor.

To counter what Rock Steady calls "the intensity of capitalism, colonialism and the dehumanization of farmworkers," the farm classifies everyone who works there as a farmer and potential member of their farmer-owned cooperative. In 2022, the farm team totals thirteen: five full coop owners, five others who share in the field work, and two who focus on administration and finances. A goal is to pay everyone a living wage and each year they get closer, sharing increased revenue among the whole crew. Everyone is on payroll, including Maggie and D, so they are eligible for Workers' Compensation, have access to paid family

leave, and, if the farm makes a profit, owners receive a portion. All farmers get a CSA share and paid vacation and paid sick leave. In 2022, the farm applied for Food Justice Certification as a way to acknowledge their commitment to farmworker justice.

As the child of a farmer, Maggie is sensitive to the many ways that growing healthy vegetables can damage physical and mental health. The farm is scaled to allow for diversified work each day. Workers are trained in food and farm safety and urged to learn new skills. The employee handbook stresses building "efficiency, speed and quality," but limits the work day to eight hours and the budget only covers a forty-hour work week. An hour lunch break and two fifteen-minute rest breaks are mandatory. There are morning check-ins, three employee evaluations a season, and the whole crew makes time for peer review and facilitated support to navigate complex issues and conflicts, when necessary.

Each year, Rock Steady has increased the percentage of their food that goes to low-income people from an initial 40 to 57 percent of the CSA shares in 2020. Funding for the lowest payments and free shares does not come out of the farmers' pockets—instead, the farm has a Food Access Fund for share subsidies and appeals repeatedly for contributions. In addition, the CSA uses a sliding scale with 30 percent of members paying the baseline, or market price point, for shares and 18 percent paying at the top end of the sliding scale, thus subsidizing those who pay less. Many members pay with SNAP/EBT.

"CSAs are so often the white mom thing and we want to tap into more diverse groups of people who are located within the local food movement," Maggie says. "Queer folks are often facing health problems like diabetes, obesity and other diet-related illnesses which means their health is compromised. We want to see how we can bring healthy food into the queer-identified community in New York City."

Community partnerships and energetic fundraising are key to share distribution. Maggie devotes a lot of time and energy determining mission alignment, share size, and content or bulk order that best fits each partner program. An important connection for relations with Millerton locals has been providing shares for the North East Community Center, which serves low-income people in several towns near the farm. The farm has working relationships with programs that cater to the needs of low-income families and especially LGBTQIA people with health problems. Some of these programs include Callen-Lorde Community Health Center, which serves ten thousand patients and offers free shares for those living with HIV; Community Access, a NYC housing non-profit; and the Free People's Market in Mt. Vernon, serving low-income Latinx and people from the African diaspora. Each arrangement is individualized and, in 2020, the farm raised enough to donate ninety free shares. Maggie negotiates collaboration with networks that provide support such as the Queer Money Project, the Queer Farmer

Network, the Northeast Queer Farmer Alliance, and the Sexual and Gender Diversity cohort of Via Campesina.

Rock Steady also provides incubation space for another remarkable project—Jalal Sabur's Sweet Freedom Farm, which grows food for incarcerated people, their families, and other people facing food insecurity in the Hudson Valley. Sabur, a prison abolitionist, racial justice leader, and member of the Soul Fire Farm Board, takes aim at the shamefully inadequate food in New York prisons, and raises funds to rent buses to bring families on prison visits.

Less unusual than the time Rock Steady devotes to social practices, but just as central to their goals, is their approach to soil care. Year by year, they manage to disturb the soil less while planting more cover crop. In 2020, they planted 90 percent of their land in cover crop. Though not certified organic, they use only organic and holistic practices, including Integrated Pest Management; substituting row cover for spraying; using cover crops organic compost and granular fertilizer; increasing pollinators through planting natives and diverse plants; and organic low-spray techniques as a last resort. In 2021, they planted flowering

Rock Steady Farm *Rock Steady farm crew bundling green onions*

Christine Hill *Dream Into Being*

perennials to support pollination. Photos of fields after an inundative rain show no standing water or signs of erosion.

In her book Resilient Agriculture, Laura Lengnick helps us understand the complexities of resilience and how to design farms that have the capacity to recover from setbacks, to respond quickly and bounce forward while contributing to the transformation of agriculture. She writes, "Diverse networks of equitable relationships build the foundation of resilience including all possible relationships—in soil, between soil, plants, animals and people, between people in community, and between communities within a region and beyond."[1]

Rock Steady Farm is an outstanding example of a resilient farm, a farm with a vision for creating "a new paradigm in a deeply unjust food system," D says.

"We want to be able to feed people who don't usually have the access to local, organic, and nutritious food. That's at the heart of what we do." ○

Note

1. Laura Lengnick. *Resilient Agriculture: Cultivating Food Systems for a Changing Climate* (Gabriola Island: New Society Publishers, 2022).

THE FARMERS ADVOCATE

Volume 28. No. 31.   TOPEKA, KANSAS.   February 15, 1906.

The Value of the Small Farm Well Tilled

REVERSE THE VALUES

In a recent review on the ever increasing value of farm lands and the difficulties encountered by the young man who would take up farming for a living, Maxwell's Talisman said:

Does not the increase in land values in this country raise a question of supreme importance with reference to the opportunities of our coming generation,—the young men who are now growing into manhood and must soon face the problem of providing a living for a family?

The price of land in all the states where agriculture has become a well-established industry, is now so high that a young man coming out of school or college, with his life and all its problems before him, cannot, in any reasonable time, in any occupation which is open to the average man, earn enough under ordinary circumstances, to buy a farm for himself, so that he may own a home. He must be either a wage-worker or a tenant farmer.

Is there not a solution of this problem which can be made to apply to every young man of average industry and capacity? And is not that solution to put the value and the power of production from the land into the boy himself, by a system of right education, rather than in the land? In other words, to make this point clear, one hundred and sixty acres of land is none too much for a man to have to furnish a good living for himself and family, under the ordinary methods of farming now prevailing in this country. But what is the purpose of working the farm? Is it not, first, that the farmer may have a home for himself and his family, and second, that he may have an income sufficient to enable them to live in comfort, with all the advantages of education which every citizen of this country craves and should have?

If that home and that income can be just as well produced from ten acres of land as from one hundred and sixty acres, the amount of money necessary to secure the acreage required is reduced from $16,000, the cost of 160 acres at $100 an acre, to $1,000, the cost of ten acres at $100 an acre.

The acreage cost may be put at $100 because, although in many places land commands a much higher price, there is still plenty of good land to be had where a young farmer could start life, for $100 an acre.

A young man with no capital except industry and ordinary capacity, can hardly hope to earn $16,000, or to in any way save it as the reward of his own labor, during the earlier years of his life. He might, if more than ordinarily industrious and economical, save enough by the time he reached middle life to buy such a farm, but he could not do so within a reasonable time after he was ready to marry and establish a home;

much less, before or at that time.

Now, instead of bringing together a sixteen thousand dollar farm and a one thousand dollar boy, suppose that we reverse the combination and put a sixteen thousand dollar boy on a one thousand dollar farm. All that is necessary to do that is to educate and train every boy who is willing to receive the training, in the public schools, from the kindergarten to and including the country college, so that he will become so skilled in the art and science of close and intensive cultivation of the soil, in the process of plant growth, in irrigation, soil culture and fertilization, in the selection of the kind of crops to grow, and in the methods processes and systems of marketing

the richer he is, and consequently, he bends all his energies to crowding out his neighbors and adding as many acres as possible to his own domain.

In the near future this greed for land will gradually fade away, and farmers will find that with less land and more cultivation, they can make more money, and that the smaller the farm the better the roads will be, the more good neighbors they will have, the better the schools and churches, the libraries and social environment, and that the greater will be the educational advantages they will be able to give to their children. With "the small farm well tended," life itself becomes a vastly more valuable and enjoyable thing than on the isolated farm,

vide a system of public school education and bring it within the reach of every boy and girl in the land, which will train every one of them so that they will know how to cultivate ten acres of land in such a way that it will yield a greater profit than a quarter-section farm ordinarily does today, and will know how to cultivate one acre of land—a home acre—in the suburbs of a city or factory town, so as to produce from it a large measure of the living for a family, notwithstanding that the head of the family, or other members of it, may be occupied in a clerical capacity elsewhere during the day or working in a factory or a mine.

"KENTUCKIANS' HOME COMING

The programme for "Home Coming Week" in Louisville, June 13 to 17, when one hundred thousand former Kentuckians are expected to go back to their native heath, is rapidly taking shape.

The first day, Wednesday, June 13, will be known as Reception and Welcome Day; the second, June 14, as Foster day; the third, June 15, an Daniel Boone Day; the fourth, June 16, as Greater Kentucky Day, and the fifth, Sunday, June 17, as "Until We Meet Again.

The address of welcome is to be delivered by Henry Watterson, and responded to by David R. Francis, of Missouri. Others on the programme are Wm. Lindsay, John G. Carlisle, John M. Harlan, Thos. T. Crittenden, Adlai E. Stevenson, etc.

FARMERS' NATIONAL CONGRESS.

At the 1905 session held in Richmond, Va., eighteen resolutions were adopted. They favor any additional legislation which may be necessary to prevent railroads from giving special favors; popular election of United States senators; a parcel post; government aid in suppressing of immigration; national aid for good roads, reciprocal tariff relations; increased appropriation for the farmers' institute division of the national department of agriculture and a recommendation that the state not already doing so make liberal appropriations for local institutes; removing the tax on alcohol used for fuel, light and mechanical purposes; a national income tax; increased appropriations for the experiment stations.

In view of the importance of these measures the Congress has opened legislative headquarters in 919 Security building in Chicago. Col. E. W. Wickey of the executive and legislative committees is the agent in charge.

Geo. M. Whitaker, P. O. Box 1332, Boston, Mass., Secretary.

Some people profit by the mistakes of others. Take, for instance, the minister who gets a fat fee at a wedding.

Put a $16,000 Boy on a $1,000 Farm Instead of a $1,000 Boy on a $16,000 Farm.

SANS SOUCI (60756) 40752, Color Black, Percheron.
Imported and Owned by the Lincoln Importing Horse Company, Lincoln, Neb.

them, that by intensive farming of a ten-acre tract, costing $1,000, your sixteen thousand dollar boy will be able to produce from ten acres a greater profit by better and more intensive methods of farming, than the average farmer now produces from one hundred and sixty acres.

It is no longer a theory,—it is an established and unquestioned fact, that this is quite practicable, and that the only element of doubt is in the farmer himself.

Of course the average farmer and land-owner imagines that the very land acreage he can get on with is a quarter section, and that the more land he has

where the owner is devoting his life to laboriously laying up money to buy out his neighbors and isolate himself still more from his fellow-man.

To carry out the plan above suggested, it is only necessary to get two ideas firmly planted in the American mind:—

That the first thing to be considered is the life we live and our relations with our fellow men, rather than the amount of money we may have in the bank or the number of acres over which we may exercise dominion.

Second, that to reconstruct our social system and solve every social and political problem which now confronts this country, nothing is necessary but to pro-

The Farmer's Advocate   *The Value of the Small Farm Well Tilled: Reverse the Values*

(image courtesy of Tom Giessel)

Christine Hadsel

ACT III SCENE III
The Grange and Grange Hall Curtains

The Grange—officially the "The National Grange of the Order of Patrons of Husbandry"—played an important role in the agricultural and cultural life of northern New England during the heyday of small family farms. Today, nearly half of the historic painted stage curtains in Maine, New Hampshire, and Vermont are in active Grange Halls or come from former Granges.

The Grange was founded in 1867 as an organization with four levels: the National Grange, State Grange, Pomonas (similar to the county structure within a state), and Local, or Subordinate, Granges. The organization was designed to serve independent farmers and farm families, people who often lived isolated lives. As a national organization, it represented its members much like labor unions were acting on behalf of industrial workers.

The Grange arrived in northern New England with four branches to their activities: politics, economics, social events, and moral reinforcement. On the political front, one of its early campaigns was to pass antitrust laws to protect farmers against excessive freight rates by railroads and exorbitant storage rates by large warehouses. Later on, the Grange advocated for free public libraries because many of its members could not afford the subscription fees of private libraries, and Rural Free Delivery, which allowed country folk to get daily newspapers and mail without having to trek miles to town. It supported the Extension Service of the United States Department of Agriculture and worked closely with 4-H clubs. To this day, the Grange advocates at national and state levels for policies that improve the lives and economic stability of the agricultural community, including bills related to transportation, farm programs, rural economic development, education, health, and safety concerns.

Apart from agricultural support, the Grange provided a way for farmers and their families to find mutual strength through ritual and social events. There were regular Saturday night dances, suppers, variety shows, bingo, and booths at county fairs where members and their children could show off produce, crafts, calves, and lambs. On the economic front, the Grange helped form agricultural cooperatives for buying supplies and selling crops. The reluctance of many insurance companies to cover farm operations also caused the Grange to establish its own insurance agency—National Grange Mutual—for the benefit of the farming community. Advertisements for that company appear on almost every Grange Hall stage curtain.

Grange Halls were invariably built with a kitchen and dining room downstairs and a large hall with a stage upstairs. Most stages were simple platforms with an arch at the front and a dressing room in a wing to one side. The space on the other side of the stage was usually fitted with storage cupboards for costumes and ceremonial items. At every meeting an upright piano was used to accompany songs such as "Stay On The Farm," "Who Will Reap?," "As We Go Forth To Labor," and old favorites like "My Darling Nellie Gray," and "The Bluebells of Scotland." The piano was also used by variety show performers for dance tunes.

Grange Hall grand drapes are generally simple in design, with only token painted drapery on each side and a central generic country scene surrounded by bands of advertisements. Therefore, along with their ceremonial function, Grange Hall grand drapes provide a way to promote local business and individuals who have paid for the scenery and probably most of the furnishings of the space.

Errol Briggs, the Vermont State Grange Master, explained the role of stage curtains in Grange Halls:

The first four degrees of Grange membership are named after the seasons of the year... so the grand drape is used to reveal tableaux that represent planting, cultivation, harvest and, for winter, a homey domestic scene. The fifth degree, 'Pomona,' is represented by high summer when fruits are beginning to ripen, and the sixth degree, 'Flora,' is represented by a rose garden. A garden scene backdrop can therefore, with a few embellishments, serve to represent several degrees.

In addition, many Grange Halls have an advertising curtain that portrays an imaginary street where each building is made up, Lego-like, of blocks of ads. The snapshot of economic life in rural America in the 1920s and 1930s provided by Grange Hall advertising curtains has become a valuable historic reference. The businesses advertised on the curtains include everything from tractor dealers to hairdressers, funeral parlors to grocery stores, and coal haulers to local "Improvement Societies." A blimp was often added as an advertisement for a hardware store or a milk company; the similarity of its shape to a milk bottle seems to have been very appealing. Almost every advertising curtain contains businesses that are still in existence, but these are heavily outnumbered by vague memories of corner stores, feed stores, and dress shops that are long gone.

In a few cases, a Grange Hall was also designed to function as a theater, probably to cater to summer visitors. These stages have multiple backdrops to complement their grand drape, sometimes created by the top scenic companies in New England. For instance, the Willow Grange in Jefferson, Maine, has the finest set of L.L. Graham Scenic Studio curtains in New England, and the set of L.J. Couch

Christine Hadsel   Former *North Anson Grange #88: Pomona* (image courtesy of Curtains Without Borders)

Christine Hadsel *Tunbridge Grange #364* (image courtesy of Curtains Without Borders)

Scenic Company curtains at the Grange in North Anson, Maine, is the most extensive example of that company's work. In North Jay, Maine, there is a complete set attributed to Harry Cochrane with an Egyptian scene on the grand drape, and at Broad Brook Grange in Guilford, Vermont, there is a set by Charles Henry featuring the chariot race from *Ben-Hur*.

The Grange in northern New England reached its peak in the years following World War I, with 35,000 members at 340 Granges in New Hampshire; 16,000 members at 112 Granges in Vermont; and 55,212 members at 419 Granges in Maine. However, by 1960, the decline of Grange membership had set in due to the shrinking number of family farms and the coming of movies and then television,

which encouraged people to stay home rather than go out to local dances or variety shows.

Today, there are 1,787 Grangers in New Hampshire in 63 Granges, 11 of which have their original painted scenery. In Vermont, there are 1,407 members in 48 Granges, of which 10 still have their curtains on-site. In Maine, there are 4,300 members in 136 Grange Halls, out of which 38 active Halls (and 18 former Halls) contain 82 painted curtains, more than half of all the historic theater curtains in Maine. Every year a few more Granges shut down and the average age of Grange members goes up. There are still Grange meetings, suppers, dances, and the occasional variety show all over New England, but many people have never heard of the Grange and have no idea of its role in

Christine Hadsel *Trenton Grange #550* (image courtesy of Curtains Without Borders)

local agricultural and cultural life for more than one hundred years.

As Maine Grange Historian Stanley Howe says, "The northern New England states are known nationally as the 'Gibraltar of the Grange'—the rock that will always survive. We have the determination to stay true to the rituals and traditions." Even so, some Granges are trying to adapt to a new agricultural world and the changed nature of entertainment. They host farmers markets and quilt shows or become the home of a contra dance group or yoga studio. But most Granges are beset by the costs of maintaining one-hundred-year-old wooden buildings with elementary septic systems, no land on which to drill a well or build a parking lot, and elderly members who can no longer fix the roof themselves. In another twenty-five years, their painted stage curtains may be the only things left. ○

A version of this profile with the same title was initially published in *Suspended Worlds: Historic Theater Scenery in Northern New England* (Godine Publishing, 2015), and was printed by permission of the author.

Zach Withers

DRYLAND, PT. II

Dryland. A farm, a ranch, a butcher shop. A new pickup truck, finally. Particles of field peas rising like smoke, lingering, falling, and cascading from the spout of the dust collector affixed to the side of the decrepit forty-year-old hammer mill we've chopped down to its bare function: rendering whole grains into powder to replace the brewers grain that thickens the thin porridge of corn mash from the all the vodka and whiskey produced in the middle Rio Grande Valley into a gruelish slop that our red and gold and mottled, waddled hogs love so well.

Rubbing hemp-infused and mentholated miracle salve onto my wrists to try to keep the early onset arthritis from crippling me. A six-year love affair with five-gallon buckets and boning knives threatening to dissolve into a nasty family divorce. Over a million pounds of produce a year without breaking a sweat. Trying not to twitch when I see food wrapped in plastic that you couldn't pay me to bring on the farm anymore. Plastic trauma stress disorder has no cure.

Hydraulic leaks and hostile bureaucrats and struggling to wrap my head around what a "chart of accounts" actually means. Fighting demons and the debilitating impacts of unaddressed mental health problems that plague my family. The purrs and clucks of the Black Spanish turkeys on windy, biting cold spring mornings. Inhaling pounds of compost while turning endless piles of shit, wasted food, and wood mulch, bathing in the sweet, steamy warm air exuding from dark humus-in-the-making and coming away three shades darker. Lunacy and love and more and more and more organic matter. It's absolutely insane the things that people throw away.

High-rise hotel conference rooms, deep in the city, where crooked union bosses call weighted votes to try to silence the dying voice of the family farm. Echo chambers of regenerative-eco-resilient–biodynamic-organic-perennial-holistic-passive-solar-agriculture fading out down the hall from the infographic-globalized-sustainable-jet-fuel crowd pandering for federal handouts to subsidize the endless overproduction of an ever-growing variety of food we grow just to burn. Presidential suites and hotel lobby bars where the lines blend and merge and boardroom politics give way to the power of camaraderie among the small crowd of hard-bit folk still scraping to make a living from the land, still trying to keep the land alive. Solidarity forever.

Cutting through the noise. That uncomfortable smirk from congressional staffers and bureaucrats when asked for something they would have to mow through an endless army of well-funded lobbyists to even bring to the floor. The bizarre sensation that comes from hearing the highest ranking federal officials tacitly admit that the power of the globalized food corporations will not be overcome through administrative reform or legislation. Good luck. Godspeed.

Ponderosa seedlings sprouting up in the ditch, foretelling the final chapters of their flat-topped elders. A young crop of purple lotus spindling up through the corpses of the cracked wood of the last generation, digging into the stream banks, bone dry for years now. For now. Our little orchard yielded a whopping seven apples last fall, one for every year we've spent pouring love and carbon into the Manzano loam. The desiccated remains of the fields of snakeweed reminding us that all things pass, this too shall pass. Succession is inevitable,

sweet, mysterious, and happening. The frogs slumbering deep in soil, under Lake Loretta where the old barrel raft lies by the dam, sun-baked and disintegrating. She'll still float though. The frogs will croak again.

The soul-sucking drain from endless arguments with scumbag lawyers over disputed access to a commercial warehouse building foolishly and spontaneously purchased—cash on the barrelhead—in the midst of a global pandemic. Part of a faltered campaign to spontaneously invent a viable local food system among a sudden collapse of the heavily consolidated corporate food supply chains. Dreaming of a meat locker, food hub, co-packing plant, cidery, distillery, Grange Hall.

Endless remodeling, scurrying to keep up with scores of local producers finding market access through our humble farm stand on an adjacent lot. Busting into USDA packing houses to save precious carcasses from rotting on the

Nina Elder *Fray V*

rail, in the midst of ego-driven managerial collapse. Table-breaking beef and singing ballads of the abattoir blues. Suet bars and tallow soaps and the mind-bending realization that the money in meat isn't in the meat, but the inbetween. The bird blocks and boot wax and boxes and boxes of raw dog food. Rendering leaf lard and simmering bone broth realizing every drop of value redeemable from the lives we take, an offering to the memory of the lives we created.

The billionaire brewery owner cut us off from the beer mash. Ten thousand pounds of high protein feed inputs a week, redirected to feed a privately owned elk herd on a two-thousand-acre enclosure where the obscenely wealthy can harvest a trophy animal whenever they want. A meager one hundred thousand

dollars if you shoot a real toad.

A neighboring town approved a megadevelopment with four thousand homes, an eighteen-hole golf course, and four commercial centers, to be built fifty feet north of our fence line in a closed door meeting—closed to everyone except the developers. They chuckled at our previous proposal to partner on a conservation development project in which we would lease the majority of the land and implement holistically planned grazing to ensure long-term water availability for a scaled-back cluster development scheme. They stopped chuckling when we pointed out that the first eight hundred townhouses on quarter-acre lots will be immediately downwind from a small commercial hog farm. We'll see who gets the last laugh.

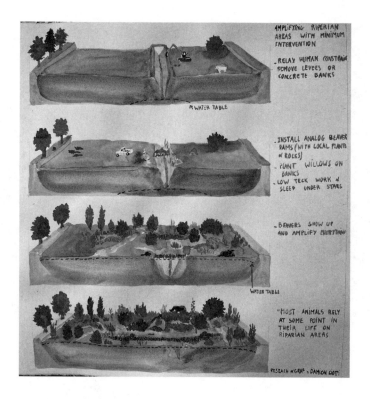

Suzanne Husky *Amplify Life Like the Beaver*
(diagram initially conceived by Michael Pollock and Damion Ciotti)

The drought is holding. The worst in a millennium, they say. It parches the soul. The rains that do come fall on sleeping seeds to be quickly carried off by harsh wind and baking dry heat. Federally recognized disaster zone, though the steady stream of transplants and climate refugees flooding in from everywhere hardly seem to notice. Our community will run out of potable water in the next few decades; this the state knows but refrains from placing any restrictions on the ever-intensifying sprawling suburban development or proposing an alternative water resource. The dry land getting drier, the semidesert with the desert's heart slowly shifting to its true nature. Harsh realities slowly sinking in. No chance in hell we turn the tide. We will be shepherds to the collapse, till one day sooner than we think, the dryland will open up and lovingly receive us back. ○

Benjamin Prostine

WHO STEALS THE COMMON FROM THE GOOSE

The Law locks up the man or woman
Who steals the goose from off the common
But let's the greater villain loose
Who steals the common from the goose

And now we follow the common goose –

– along the frowning fences of corn fields,
beneath suburbia's gates and cameras,
past every flame twirling guard of Eden –

To the squat in the old slaughterhouse,
to the lab where they probe the wheat for profit-tolerant genes,
off the highway, down by the park, below the bridge –

By the parking lot in blackberry canes,
in hen-of-the-woods grown for no one on the trees,
past campus squares and bulldozed earthen mounds,
camps of detainment, cop shops,
razor wire prison vortex and tenements on bones –

I follow the common goose

I carry the ancestral rhyme

I ask, to whoever will listen,
who steals the common from the goose?
who's law locks the man, woman, child, hen, hog, anyone up?

        Friends! Comrades! Memory makers! Ghosts!
This is not our law
      nor our justice

       The greater villain is still loose

       Who kills the common

        Who kills the ocean

         Who kills the goose

       It's later than you think
       Honk! Nip! Hiss! Piss!

          all our dreams are illegal

Alright alright, everyone's attention please,
one more time, come on, all together now,
a one and a two and a three:

omnia sunt communia

       For the quartered and drawn
       for the stocked and chained
       for every witch hunted and burned

        omnia sunt communia

From Abu Ghraib to Guantanamo
from Rikers to Waupun
New Lisbon to San Quentin

omnia sunt communia

for the motto with the rainbow
the covenant with the cosmos

omnia sunt communia

like voices out of a whirlwind
out of tear gas, out of hurricanes, out of love

omnia sunt communia

all things held in common

over and over again

here, there, every day

from the plains to the plates,
from seeds to water to air –

Mother Goose hears her children's blues
Mother Goose flies across the world and hears those blues
Says "Down with the fence! down with the wall! let loose the common
now! now! now!"

one more time

omnia sunt communia

Kim Kelly

GROWING OKRA (AND BEANS AND TOMATOES...) FROM CONCRETE

An Interview with Farmer Shawn of Life Do Grow Farm

In the heart of North Philadelphia, nestled within the poorest zip code in the poorest major city in America, something green is growing.[1] If an unsuspecting visitor who happens to be walking up W. Dauphin Street's cracked sidewalk turns on just the right corner, they'll be greeted by an oasis of foliage, trees, flowers, and a greenhouse overflowing with produce, all springing up from the concrete in a riot of life and color. They'll see a team of young Black and Brown people carefully tending to the soil, sharing cooking tips with elders, and teaching neighborhood children how to grow their own tomatoes.

This lovingly tended green space is called Life Do Grow Farm, and is the anchor project of Urban Creators, a nonprofit community organization founded by local college students in 2010 that now provides fresh, healthy organic produce to hundreds of North Philly families.

For the past dozen years, those involved have been working to educate, organize, and inspire their fellow North Philadelphians, particularly the youth, and Life Do Grow has served as a centerstone for their work. Not only is it a working three-acre organic farm with a seasonal community-supported agriculture (CSA) program and weekly produce markets,

Life Do Grow also serves as an employment center for students and people returning from incarceration; is the site of numerous community events, from hip-hop concerts to West African cooking classes; and regularly hosts local Black makers, artists, and volunteer groups. Social justice—particularly racial and economic justice—are core elements of Urban Creators' message, and, as the team has expanded, so has the scope of the group's mission.

"Over the last two years, Urban Creators has begun a transformation from an organization focused on urban sustainability to one focused on food sovereignty," Director of Operations Elizabeth Okero wrote in an 2022 op-ed for the Philadelphia Inquirer.[2] "Food sovereignty means having the right to foods that are culturally appropriate for you. It means healthy and ecologically sustainable foods, and the right to define your own food and agriculture system."

That guiding principle animates everything they do on the farm and out in the community, and one of the team's most energetic proponents is a thirty-four-year-old Black man who's known to everyone as Farmer Shawn. Though he was born and raised in Brooklyn, the lifelong punk rocker has made a new home in South

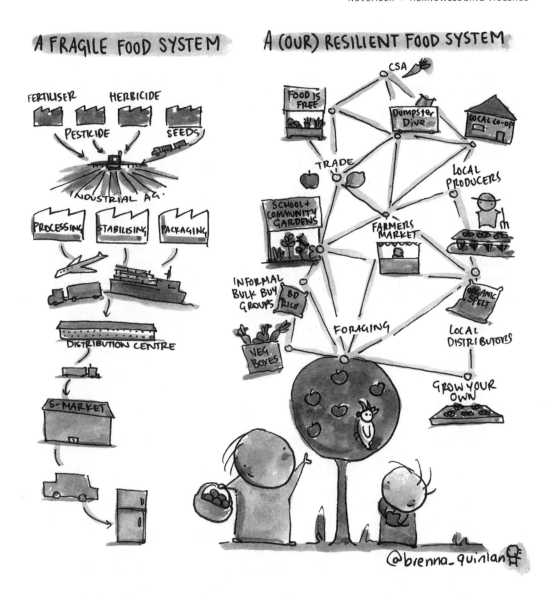

Philadelphia (with, well, me: full disclosure, Shawn is my longtime partner). He has become a key member of the Life Do Grow team as the Farm Manager, and can usually be found doing any number of extra projects on top of his

regular duties, like building out new structures, making compost, and teaching self-defense classes to local youth. This is his life's work, and, as our bursting bookshelves can attest, he is always pushing himself to learn more about organic agriculture, to share more of that essential knowledge with the people, and to become the best farmer he can possibly be.

On a hot August day, we sat down on our couch to talk more about the work he and his coworkers are doing with Urban Creators, and why urban farming and food sovereignty are so important for marginalized communities.

How did you start working in urban agriculture? I know you landed at Garden of Eve Organic Farm and Market in Long Island when you were still a teenager.

My parents were weirdo hippies and they really cared about their kids eating healthy and were into their organic stuff before it was super cool. They actually used to work extra hours so they could afford organic vegetables. We had a little house with a little garden. Being raised by people that were also pro-union and had a picture of [abolitionist] John Brown on our wall, the idea of food justice came naturally to me. If you mix gardening with people fighting for their lives, that's where you get food justice—it's a combination of fighting for one's sovereignty for not only oneself, but also for one's community.

I started working on that farm after I found a job that mixed construction and farming. Because of the construction job, I knew I could do repetitive things in either really cold or really hot temperatures, and farming was just something that I wanted to learn. Also as a young Black man, I looked up to organizations like the Brown Berets and the Black Panthers and how they wanted to improve their community; I wanted to do that—and that all starts with food.

After that, you started working for Grow NYC, an urban sustainability nonprofit in New York City, where you built and maintained hundreds of community gardens throughout the five boroughs and got hands-on experience teaching people how to grow food. You were there for almost ten years—and then what happened?

My sweetheart lived in Philly, so I was going over there part-time. Then COVID-19 happened, and then the shutdowns happened, so I was kind of like, well, I guess this is my home now!

In 2020, I got a job at the Weavers Way Co-op's Henry Got Crops Farm and worked there throughout the first year of the pandemic. It was cool; it was not a lot of pay, but they do a really good job at growing. I worked at a grocery store for a few months, and then found the listing for Urban Creators on Good Food Jobs. I went through three rounds of interviews, and I guess they liked me, because I'm now in my second year as farm manager.

I do the crop planning, the planting, sending out seed orders, managing soil health, pest management—pretty much like anything that is involved with either stewarding the land, improving the soil, or harvesting. From growing the crops from seed to harvesting and everything in between.

And aside from your hands-on experience at work, you're completely self-taught when it comes to your knowledge of organic farming, right?

Yeah, having a lot of jobs in urban agriculture and horticulture helps! But seriously, I'm just a super big nerd about reading anything about [organic farming]. I go to used bookstores and just pick up old books because there's really good information in there, and then YouTube is amazing. When I started doing this stuff, there weren't as many people on there who were into farming, but it's got super trendy, which actually helped me a lot. And right now I'm actually lucky enough to actually put that information into practice. I can try different things and see how they work.

I also look up to other Black farmers like Leah Penniman. If you talk about Black farming, you need to mention her name. She's been a leading force for Black farmers of all genders and is awesome. Culinary historian Michael Twitty is amazing, too. Recently I've been getting into cooking in terms of how it relates to farming, because what's the point of growing food if it doesn't taste good?

Urban Creators is a predominantly Black organization, working and serving a predominantly Black community in a very Black city. How has working in this specific environment differed from other agricultural jobs you've held? What kind of impact has it had on you as a Black farmer?

At almost all the places I've worked, I have been the only Black person. Being around people that are deeply melanated like myself, there's a level of comfortability. It feels really good to be around people that have had the same experiences with oppression, but you can also just joke about stuff and they know exactly what you're talking about. They have been through it too.

The mission becomes super focused and singular. We're in a Black community, we are Black people ourselves, and we are here for the people. We get to feed them really good food that tastes good, make sure they're getting proper nutrition. We teach them how to grow their own food and be self-sufficient, so they're not as beholden to a system that doesn't really care about us. This is the goal that we're setting, and we're gonna get there, and to hell with anyone that's going to get in our way.

Why is it so important for Black folks, especially poor Black folks in the city, to be able to grow their own food and share that with their neighborhood?

One word comes to my mind: autonomy, the ability to operate without any hierarchical or government structure. It's really important that Black folks from every walk of life learn how to feed themselves, because that is how we take the power back from the systems of oppression that feed us government cheese and bad milk and all that stuff. It's the ability of regaining our control over our own destinies; as some creep once said, "Who controls the food supply controls the people," and if we learn how to make food ourselves, we can control ourselves and we can influence our communities to be more autonomous.

At Urban Creators, we have a lot of irons in the fire, and we're lucky to be able to share our knowledge and our practices with other folks. As a big dream, I would love to see something like Urban Creators within eight blocks of every community—because it's super localized, you wouldn't have to go out of your way to get proper food in areas struggling under food apartheid. Expanding the knowledge so people have the confidence to start something on their own. ○

Notes

1. "Map: Income and Poverty in the Philadelphia Region," The Philadelphia Inquirer. 2018. *media.philly.com/storage/special_projects/income_and_poverty_in_the_philadelphia_region.html*.

2. Elizabeth Okero, "At Life Do Grow Farm, North Philly Residents Plant the Seeds of a Sovereign Future," *The Philadelphia Inquirer*, January 1, 2022.

335

Erwin Driessens & Maria Verstappen   *Morphoteque #9*

Zohar Gitlis

A PRAYER FOR JEWISH FARMERS

Stung by centuries of old taunt that. . .the Jews of Eastern
Europe did little to build up the lands of their birth, a back-to-
the-land movement for the Jewish masses in Poland, Lithuania,
Latvia and elsewhere. . .is now in full swing. . .Forests never
touched by the toll of man now resound to the woodman's axe.
Ground hitherto unfurrowed by a plough is being broken to
sustain hundreds of Jewish families who have torn themselves
from crowded ghettos for the adventure. . .

—*The New York Times*, February 1, 1925.[1]

On my twenty-sixth birthday, I began working at a farm in northern New York that would become my home. We harvested and peeled leeks in the damp, biting December cold. As the months passed, I became obsessed with the way it felt to eat food that I had labored to produce and the discovery of strength in my body that I had doubted it was capable of.

A visitor once described our farm crew as "unbelievably good-looking." The comment stayed with me because it pulled uncomfortably on my own desire to cultivate a body and persona that fit an imagined ideal of a young farmer. I was swept up in the romance and eroticism of the farm: muscles that gleam with sweat, hairy armpits that smell like soil and pheromones, long unkempt hair swinging to the rhythms of a rusty tractor, and kisses snuck between rows of sweet corn.

My ancestors were impoverished Jews trapped in ghettos on the borderlands of tsarist Russia, the pale of settlement. When I imagine my ancestors fleeing pogroms and anti-Semi-tism, I see the farm's opposite: bodies weighed down, slumped, pale, and prematurely aged from destitution and displacement. The young Jews of the Ashkenazi back-to-the-land movement, described above by the NYTimes in 1925, brought young Jews to colonies in Ukraine and Crimea, the Canadian prairie, New York's Catskills, South American ranches, and Palestinian swamps. These settlers, my ancestors likely among them, were to become the "New Jews."[2,3]

These agricultural settlers emancipated themselves from anti-Semitic Europe and reconnected to the land, rather than retreating into traditional religious study and accepting the taunts and slurs of their oppressors. Land, and the idealized vision of being capable of manipulating land became a tool to seize and embody one's own freedom. Muscles tightened, backs straightened, the weary and wandering Jews transformed into New Jews.

For thousands of years, Jews have lived on borderlands, at the behest of empires, rejected

Zoltan Kluger *Members of Kibbutz Manara planting fruit trees on the slope* (image courtesy of National Photo Collection of Israel, Photography Department Government Press Office)

by nation-states that have come and gone. We've been the subject of the "Jewish Question,"[4,5] a query left eternally unanswered that seeks a solution to our wandering, our perpetual mismatch with the Christian desire for homogenous society. Over time, Ashkenazi Jews became pawns in Western schemes to whiten the American colonies and the Middle East: determined white enough for distant lands and given acreage to displace the indigenous, darker others.

Today, my Jewish comrades try to make sense of these early 20th century back-to-the-landers, to read them in the context of the Torah's utopian agriculturalism.[6,7,8] A utopianism that emerges from the ancient Jewish practices of shmita and yovel, the Jewish Left's favorite biblical mandates to reset ownership and let land lie fallow in cycles of seven and fifty years. The motivations of our ancestors remain unclear to us, though, opaque as our contemporary motivations. A jumble of desire for freedom, romance, and spiritual wholeness; entangled with imperial impulse to win freedom through conquest; and spiritual wholeness through religious domination. To untangle these desires is an impossible, yet necessary, task.

Here, then, is my prayer for us, the Jewish farmers, and all farmers:

Notes

1. The New York Times. "Jews In Eastern Europe Join Back-To-The-Land Movement." February 1, 1925. *timesmachine.nytimes.com/timesmachine/1925/02/01/104165403.html?pageNumber=178*.

2. "New Jew" is a term commonly used to describe Zionist agricultural settlers in Palestine, and their Palestine-born children, prior to the 1948 war that displaced Indegenous Arab populations and the creation of the Israeli state. The term connotes a shift in temperament and physique that settlers believed would come with emancipation and independence. Israeli sociologist Oz Almog describes this generation: "born and bred on his own land, free of the inhibitions and superstitions of earlier ages; even his physique was superior to that of his cousins in the old country."

3. Almog, Oz. *The Sabra: The Creation of the New Jew.* Electronic resource. The S. Mark Taper Foundation Imprint in Jewish Studies. Berkeley: University of California Press, 2000. *columbia.edu/cgi-bin/cul/resolve?clio10313196.001*.

4. The "Jewish Question" refers to the political debate over the place of Jews in European society during the 19th

When diaspora begets diaspora,
displaced peoples displace people,
and agriculture becomes a tool of settlement,

I pray for us to see the ripples of our own transformations.
When the New Jew became an agent of the New Empire, our yearning for freedom and transformation is weaponized against justice.

When land is transformed from swamp, from forest, from prairie, into cultivated fields,
sun rays into calories,
Old Jews into New Jews,
the morality of imagined utopias gets murky.

I pray we see the consequences of our transformations, which we are forever obligated to address. No transformation, regardless of our motivation, is without consequence. I pray we have the courage to understand that there will always be consequences, and thus our visions of freedom must always evolve and expand.

I pray for us to imagine beyond what we've ever imagined before.
And, then, to understand that true freedom is beyond even these wildest imaginings.
Yet we get closer when we continue to imagine,
honoring our ancestors not by holding tight to their visions, but expanding the boundaries of possibility.

I pray that we imagine beyond land as a tool to seize power and individual freedoms, and toward land as a partner in pushing the natural and social worlds we live in
closer and closer to justice. ○

and 20th centuries. Although the term is traced to the onset of modernity, it often implies the eternal minority status of Jews in European society.

5. Philippe Hoffmann (Paris), Christhard (Bergen Hoffmann, and Doron (Jerusalem RWG) Mendels. "Judaism." *Brill's New Pauly*, October 1, 2006. *reference-works.brillonline.com/entries/brill-s-new-pauly/*-e1407860*.

6. Shmita and Yovel (Jubilee) refer to the seven and fifty year cycles of agricultural reset outlined in Leviticus. The Shmita, or seventh year, is set aside as a sabbath during which agricultural land is to lay fallow. The Yovel, or fiftieth year, is both a Shmita year, and a year when all slaves are to be set free and land returned to its original owners.

7. Jewish Farmer Network. "What Is

Shmita?" *jewishfarmernetwork.org/whatisshmita*.

8. Orthodox Union. "Shemitah and Yovel Sabbatical and Jubilee Years." *ou.org/judaism-101/resources/shemitah-2*.

Support the Farm Workers

can <u>your</u> family live on less than $1800 a year ?

DON'T BUY CALIFORNIA GRAPES

CALIFORNIA TABLE GRAPES WERE PICKED BY PEOPLE WORKING 10 HOURS A DAY IN THE FIELDS WITH NO BREAKS AND NO TOILETS. EVEN IF EVERYONE IN THE FAMILY WORKS, THE FAMILY CAN ONLY EARN $1600 - $1800 A YEAR. THE FAMILY IS FORCED TO GO ON WELFARE WHILE THE GROWERS EARN MILLIONS. TO HELP FARM WORKERS GET OFF WELFARE AND GET A LIVING WAGE AND DECENT WORKING CONDITIONS THROUGH RECOGNITION OF THEIR UNION, DON'T BUY CALIFORNIA TABLE GRAPES.

Talk To Your Grocer

Ask him not to sell any California Grapes until the Grape Growers sign contracts with Cesar Chavez's United Farm Workers Organizing Committee, AFL-CIO.

N.A.A.C.P.

948 Market Street - Suite 703
San Francisco, California 94102
(415) 986-6992

National Association for the Advancement of Colored People, Los Angeles, California Branch

Support the Farm Workers (image courtesy of University of California Berkeley, Bancroft Library via California Digital Library)

Abdullah Shihipar

ROOTS OF RESISTANCE

The Legacy of Nagi Daifullah

On an early August morning fifty years ago, a group of striking grape workers were at the Smokehouse Café in Lamont, California, after a day of walking the picket lines. One of those people was a twenty-four-year-old Yemeni organizer and worker named Nagi Daifullah.

The year was 1973, and that summer grape growers across the Coachella and San Joaquin valleys launched a strike that would last for three months and see the state crackdown on the strikers through mass arrests and violence. It was just after one o'clock in the morning. Under the glow of the moonlight, the workers were gathered at the café. Perhaps they were trading stories about the day's pickets over coffee and cigarettes, perhaps they talked about their family and friends—people they left behind for the hot days and intense labor of the farm. Whatever they were talking about, the early morning hours of August 15 live in infamy in the history of the United Farm Workers, for what happened next.

A police car from the Kern County Sheriff's office pulled up and began harassing the workers as they had done all summer long. An officer, Deputy Gilbert Cooper, attempted to arrest one of the workers. In the process, other workers, including Nagi, began to protest. Deputy Cooper then turned his attention to

Nagi and pursued him on foot, striking him with a metal flashlight on the back of his head. After Nagi was hit, the deputies dragged him back to their police car and left him in a gutter, bleeding from his head. Nagi had sustained damage to his spinal cord and, despite pleas from other farmworkers to call an ambulance, none was called. Nagi Daifullah died that morning, in a gutter, beside a police car.

We know these grim details because the UFW documented Nagi's death in a pamphlet about people who died for the cause.[1] Listed alongside Nagi are the names of Nan Freeman, a student killed in an accident with a truck at a picket in 1972, and Juan De La Cruz, a farmworker shot and killed on the picket line by a scab worker two days after Nagi. Alongside details of Nagi's gruesome death, we hear about his dedication to the cause; his ability to speak Arabic, English, and Spanish and assist with interpretation while organizing other Yemeni, Arabic and Spanish-speaking workers. He was a small young man, weighing no more than one hundred pounds and standing five feet tall. Like many migrant workers who come to America to work in the fields, he would send money back to his family in Yemen. Days after he was killed, thousands came out to pay their respects to the young man—the funeral procession led by none other than Cesar Chavez himself—before his body was flown back to Yemen for the burial.

Those who knew Nagi talked about his leadership in the strike, his ability to speak multiple languages bridging gaps between workers of different ethnicities and languages.[2] Ahmed Yahya Mushreh, a former farmworker, told *SFGATE* in 2002 that Nagi would encourage other immigrant workers by reminding them that they had the right to organize and the right to fight for their rights. "This is democracy, and if you want your rights, this is how you do it. You fight for your rights. This is the United States," he would say.[3]

Like myself, Nagi was a Muslim and, like me, he became acquainted with the labor movement in his twenties. He was a martyr, a hero who died in the name of defending workers. Martyrs exist in our imagination as revered figures, but they do not ask or plan to die. They seldomly ask or seek attention for the work they do. Organizing, by definition, is work that is part of a collective; you do the work, but you are successful because countless others are doing similar work, too. Martyrs like Nagi become defined by the circumstances of their death. It is the death, the act of paying the ultimate price, that elevates them into our histories and our consciousness.

There is a direct line between the deaths of Nagi and other farmworkers who died in the strikes of the 1970s and the workers who work in America's food chain today. Today, America's food systems run at a backbreaking pace, killing and maiming workers at all levels of the food chain. Workers who pick produce and process meat, as well as the warehouse workers who pack and prepare food for shipment, are all at risk for exploitation that puts their lives at risk. A dangerous job became even more

fraught with the onslaught of a global pandemic, when these workers were deemed "essential" and had to face daily exposure to a virus.

COVID-19 spread like wildfire across meatpacking plants. With disregard to wellbeing of workers, leadership at these plants worked with the Trump administration to stay open during the worst of the pandemic.[4] At least fifty-nine thousand workers caught the virus and 269 died as of October 2021.[5] While the virus spread, managers at one Tyson plant in Iowa took bets on how many workers would get COVID-19.[6] For those who work out in the fields, COVID-19 exposure is not the only health issue that the occupation presents: as the climate continues to warm, heat related illness poses a threat. Three hundred eighty-four workers have died in the past decade of heat related causes, and farm workers are thirty-five times more likely to die of heat-related illness.[7,8] It is why the UFW continues to lobby for the passage of a heat bill that would extend heat-related labor protections to farmworkers, and why the Occupational Safety and Health Administration, or OSHA, is working on a labor standard that would address heat.[9,10]

The pandemic has created nightmarish conditions for workers; there have been few increases in pay and temporary measures like hazard pay and extended sick leave have expired. With a stated "return to normal," workers are working with few protections from the virus, which is still killing hundreds of people a day as of summer 2022. We don't want more workers to die in order for there to be progress; we want workers to live to see the fruits of their labor and organizing. But just as the pandemic has harmed, it has also created conditions ripe for organizing, as workers across industries hold rolling strikes, work

The United Farm Worker (UFW) Collection
The funeral procession of Nagi Daifullah.
Over seven thousand mourners march with the casket of the slain farm worker Nagi Daifullah en route to his funeral at the Forty Acres, Delano, California, August 1973. Dailfullah was a twenty-four-year-old picket captain of Yemeni descent who died after being struck in the skull by a police flashlight. Cesar Chavez stands to the left of the American flag, Dolores Huerta stands to the right (in poncho), August 1973. (image courtesy of Walter P. Reuther Library, Archives of Labor and Urban Affairs, Wayne State University.)

stoppages, and push for unionization.

Fatalities like those of Nagi's should be seen as part of the regular death-making of the food system, not as an exceptional event. Killing and maiming workers who stand up for their rights enables the system to further kill and maim without impunity on a regular basis. When these workers are killed their colleagues are robbed of comrades and their communities are robbed of people with knowledge and histories. Death is used to intimidate other workers into silence—but in making martyrs of worker-organizers, they have only furthered worker-led resistance. The strike continued after Nagi's death, and once again a boycott of offending grape producers was organized. In 1975, farmworkers won the right to collectively bargain in the state of California.[11]

Today, I see reminders of Nagi's work everywhere: in the strike organized by Yemeni bodega owners in response to the Muslim ban, in the walkouts led by Somali workers at

Amazon facilities in Minneapolis demanding time off for Eid. Being Muslim and a worker in the United States is to face dual attacks on your identity and your livelihood; Nagi's legacy reminds us that these things are not separate from each other. Nagi used his immigrant background and abilities to improve working conditions for all workers, to build coalitions between workers of different ethnicities.[12, 13]

As I write this, I imagine Nagi today had he not been brutally murdered. He would have been in his early seventies, maybe we would have met at a picket line somewhere, maybe we might have traded generational stories over coffee. Maybe he could have given me advice as an elder. But I will never know Nagi as an elder—because he was martyred—and those of us who are born decades later are robbed of learning from our martyrs as they lived, so we must suffice for learning from them in death. But Islam teaches us that death is not the end of our story.

The Prophet (peace be upon him) tells us that we should be in this life as though we are strangers passing by. When we hear of someone's death, we say "to God we belong and to him we return," an ever-present reminder of the fragile nature of our own lives. We must resist the process of death- making and make life better for others, not because we do not believe in the finality of death but because we know there is life after it and to get there, we must earn our place. One day, god willing, I may meet Nagi in the afterlife and say *salams*, the Islamic greeting of peace, to him.

Until then, his legacy serves as a reminder of what we must do: fight for the rights of ourselves and others, work to bring people together, and push for dignity and respect for all people; with the reminder that any of these actions could potentially be our last. ○

Notes

1. United Farm Workers, "UFW Martyrs," *ufw.org/pdf/Martyrs.pdf.*

2. Marc Grossman, "Remarks by Marc Grossman for the United Farm Workers and Cesar Chavez Foundation at Observance Honoring Nagi Daifallah, September 28, 2017, U.C. Berkeley," *United Farm Workers*, September 29, 2017.

3. Tyche Hendricks, "Legacy of Yemeni Immigrant Lives on Among Union Janitors / Farmworkers Organizer to Be Honored in SF," *SFGATE*, August 16, 2002.

4. Kate Gibson and Irina Ivanova, "Meatpackers Lobbied to Stay Open as COVID-19 Spread, Congressional Probe Finds," *CBS News*, May 12, 2022.

5. Josh Funk, "Report: At least 59,000 Meat Workers Caught COVID, 269 Died," *AP News*, October 27, 2021.

6. Laurel Wamsley, "Tyson Foods Fires 7 Plant Managers Over Betting Ring On Workers Getting COVID-19," *NPR*, December 16, 2020.

7. Robert Benincasa, Julia Shipley, Brian Edwards, David Nickerson, Stella M. Chávez, and Cheryl W. Thompson, "Heat Exposure in the US Has Led to Hundreds of Worker Deaths Since 2010," *NPR*, August 17, 2021.

8. Diane M. Gubernot, G. B. Anderson, and Katherine L. Hunting, "Characterizing Occupational Heat-Related Mortality in the United States, 2000–2010: An Analysis Using the Census of Fatal Occupational Injuries Database." *American Journal of Industrial Medicine* 58, no. 2 (February 2015): 203–211.

9. Jocelyn Sherman and Rob Bonta, "UFW Foundation & United Farm Workers Applaud Bill Protecting All US Farm and Other Workers from Heat," *United Farm Workers*, March 26, 2021.

10. "Heat Injury and Illness Prevention in Outdoor and Indoor Work Settings Rulemaking, Occupational Safety and Health Administration."

11. "UFW Chronology" United Farm Workers, 2022. 12. Annie Riley, "Nagi Daifallah Tribute," *Yemeni American Merchants Association*, 2022.

12. "Shakopee Amazon Workers Protest Wages, Lack of Time for Eid Holiday," *KSTP*, May 2, 2022.

Erwin Driessens and Maria Verstappen   *Morphoteque #8*

Tatiana Dolgushina

POTATOES WITH THE SKIN ON

potatoes with the skin on
and all other things I'm
afraid to say to my mother

you know how the soil
ruptures beneath your feet
with the feces of pigs and

ungulates. how can you eat
that trash, she would say,
so I'm afraid to speak

around her Soviet knowledge
there was only one cow
at the time of invasion,

and they took it so that my
lineage would starve, and
what else did they take

more than the cow could
provide, between the ribs
what else did they take, the

young men with guns, who
only knew their country home,
as if blinded by the dry snow

of Siberian winter, where
they would later die, with
where millions of others

would sink into the earth,
inside the cracks with the
potato skins, inside that

bloody soil harnessing the
feed, inside that foiled time
of century we're not allowed

to talk about, there where
young boys have gone to
die because they sought what

death's supposed to feel,
what's now supposed to be
now that we're left, just us

where are we supposed
to go, how can you eat that
trash, my mother says to

me, after we've both lost it
        our homeland, after
we have both lost the earth

Repose

Recovery,

rest, facing death,

eco-anxiety

Jackie Summell *Solitary Gardens*

Michele Scott

A SPACE TO HEAL

For thirty years, I woke up in a small room with seven other women. "Home" was a metal bunk bed in a room so cramped we could barely walk a few feet.

Incarceration is filled with moments that dig deep into your soul. Finding a way to survive those moments—thrive in them, even—is the epitome of resilience. Initially, I started gardening as a way to deal with my time. Kneeling in the dirt, pulling weeds, and contemplating another turn of the seasons, I discovered healing. In my first years, the brief freedom from my prison cell sustained me and gave solace from my pain. Outside, I felt the illusion of being free. The garden provided the balance that I needed to survive my incarceration.

What saved me in those decades of confinement were little patches of dirt in front of my unit. Pulling weeds became my refuge, my sanity, my healing. I found that more than just plants emerged with the little sprouts of flowers I grew: in tandem, I learned the value of life lessons. The gently nodding blossoms became a balm to my soul, and a means to fight off the isolation and sadness that permeates daily life in prison.

What emerged from my efforts to cultivate plants in front of my housing unit was a connection to my prison community. Initially, those precious free moments outside were solely for my enjoyment. Over time, the journey that kept me going became a pathway to discovery. Once I looked past the hardened attitudes and bluster necessary to survive in prison, I could see we were not so different from one another. The garden became a bridge to individuals I wouldn't have gotten to know otherwise and gave us a shared language of belonging and nature. Through these conversations I discovered my community, my people.

Others were pulled towards the small green areas I tended, and wanted to be part of this as well. With an invitation to sit down and talk as we pulled weeds, doors opened and mentorships formed as we listened and shared space. I realized our humanity surpasses social barriers. We all wanted to let down our walls and be ourselves. We all had grandmothers who at one time had us in the garden, or planting geraniums.

Space was allocated for a formal, landscaped garden, which we actively had a hand in designing, building, and installing. Diverse people within the prison came together to learn about drought-tolerant plants, the healing properties of natives, and the importance of soil balance. Our shared curiosity and excitement made these sessions the highlight of each week. The energy uplifted our spirits and carried many of us through the following days. In the prison garden, we were people with one purpose, permitted to explore and create. It was sustenance, and it was beautiful.

Now, I am home, and feel isolated and lonely, separated from my garden and the connections that sustained me while inside. I have escaped the persecution of being an "inmate" and fled the adverse environment that is incarceration. But I return to a different world. Many things are strange to me, like common language: ethernet ports, retweets, and links. There are subtleties of tap to pay, sensor toilets and sinks, and self-checkout at grocery stores. Considering those challenges

while navigating the profoundly unfamiliar language of technology, at times I feel like a foreigner in my own country.

Adjusting to the world while trying to find my place in it is exceptionally stressful, and I have wondered how to gain back the sense of belonging I found in our garden at the prison. The opportunity has emerged to work on a community garden project with a liaison from the San Diego District Attorney's Office. Presented with a blank canvas, I can tailor it to reflect the needs of the local community and formerly incarcerated people.

There are parallels between formerly incarcerated individuals and refugees, and this relationship has emerged as the core focus of the project. Both communities experience a sense of isolation and are confined by invisible barriers. Both live in a society that is unfamiliar, that we don't feel we belong to. We are two groups of human beings who need to heal after being removed from the worlds we knew, and face starting over in a culture disinclined

to hold the door open for our entry. A space to share the universal language of gardening is more than an oasis—it represents a solution.

What better way to ease the uncertainties of things once familiar and now changed than the green, healing space of a community garden that is both a buffer and the balm? With its mission to provide for these groups a green, healing space; opportunities to grow food; and connection to their communities, I see the possibilities of this project and space, where a diverse spectrum of participants can share the similarities of their respective journeys of discovery and find a place to start over. ○

Eva Parr

ODE TO COVER CROP

Thus when her enemies came to kill her, she knew more about surviving than they did about killing.
> —Octavia E. Butler, *Wild Seed*

First frost came early. Scary early.
A quiet blanket of ice enveloped the farm; death devoured fields of flowers
and ripe fruit.
I sighed with relief, despite myself.
Exhaustion weighed heavily on my muscles and mind. The season had
seen so much of us. Summer has a way of cracking you open and exposing
all the raw parts of you that hurt to look at, a seed shedding chaff.
We mowed all the tomatoes and reveled at how so much can become so
little, so quickly.

The day we sowed the first seeds was a frenzied race with the rain.
We rushed around
Laughing, tilling, raking and shaping beds, tending to decay—
This crew was fairly fresh, only two months together.
But two months of life and death is an eternity.

We all shoved our arms elbow deep into the sack of seeds in childlike awe,
pulling out handfuls of vetch, clover, wheat, field peas and fava beans. We
spread the seed by hand over the freshly washed canvas of our upper field.

That evening, as I walked up from the bay after sunset, the first drops hit
my cheeks. I tilted my head back, hoping to wash the last two years of fire,
smoke, heat, drought, death, grief, and loss from my skin.

I breathe in.
Time seems to widen.

The storm lasted seven days, flooded all the roads, and broke October rainfall records. The spillways spilled. The seeds swelled as we slopped around the mud, tucking strawberries and garlic into their wet winter beds.

Weeks later, when the sun finally emerges the seeds stretch out and bask as we do. The oats come first, reaching straight up to the light. The others sprawl across the soil surface, swallowing the earth in vivacious green tendrils.
Sparrows (debatably our worst farm pest) emerge to eat the tender young seedlings, growing into their plump winter bodies.

I smile, despite myself.

The work is hard. The days are long.
We who work closely with the land cannot ignore the grief of climate collapse.
Every year another record-breaking catastrophe. Every year uncertainty.
How many more friends' homes and farms will burn down? Will the hermit thrush return this winter? Will the spring dry up?
Every year wondering how we can go on like this.

I am comforted by the seedlings. I see resilience in the way our winter squash grows thick and beautiful skin to protect and preserve its sweetness. I relish in the song of the golden-crowned sparrow in the fall, by the smiles of my crew, my landmates, and dear friends, lit up by the sun.

We have to remember how to be in relationship with one another. We have to plant some seeds that we will not harvest for profit. We need reciprocity with the living world. Our survival depends on it.

I lie in this field of tall grass and legumes, during the driest February on record. My dog sniffs around me. Frogs croak above my head. I nibble on the fava greens and let the sun warm my wounds. I recommit to another year on the land. Another year of heartbreak, awe, sweat, flowers, frustration, belly laughter, and calloused hands. Another year of hardening and softening, adapting, learning, loosening the grip, and falling in love over and over and over again.

Falipa Lilias *Drama in the Cover Crop*

Taylor Hanigosky

AUTUMN OLIVE WITH THE SILVER LEAVES

The day the autumn olive leaves fell, I felt lightheaded and exposed. It was late December. For three months, the procession of naked release was steady and rhythmic. The poplar and black walnut, to the oaks and maples, then the low-growing and spindly sassafras and spicebush. When I peered into our woodland edge beside a small clearing, I saw a choreographed undressing from tip to soil. But the day the autumn olive leaves fell, I felt pushed out of the nest. I felt like the drivers of the cars on the newly visible road could see my thoughts. I spent the last year untangling opportunistic vines, trying to make a space for myself here. Yet, the relative spaciousness of the thicket in winter illuminated the knowing that I'd grown to fit the openings already there, curving and winding and triple-knotted in undergrowth that might be called feral, or untended, or neglected.

I'd grown enmeshed with the openly admonished and so-called invasive species. Eleagnus umbellata. I drank her tea, I ate her food, I wove her limbs, and I wore her shimmering silver shawl. We were fast friends and the friendship is intimate. But intimacy unveils complexity. Autumn olive was introduced to the United States by the federal government in the 1800s for erosion control and rapid-grow-

ing wildlife habitat in areas impacted by extractive industry.[1] Autumn olive's nitrogen-fixing tendencies permit it to flourish in depleted soils where little else can grow with such vigor. Intentional plantings continued into the 1990s, and autumn olive proliferated from Georgia to Maine. The shrub quickly crept into woodland edges and now healthy swaths of forest, outcompeting native species for available space, resources and light.

In her growth habit, autumn olive is a colonizer. She grabs at land and pushes others out rapidly, benefitting from an unbalanced and foreign ecosystem that hasn't yet had the time to evolve a control method. Yet, in her social and cultural context, she is enslaved labor stolen from her native lands to clean up someone else's mess. Can we really blame autumn olive for what she's become? My own ancestry is delineated from colonizers in North America who have directly performed more violence to this ecosystem—through land-theft, clear-cutting, swamp-draining, mining, polluting, and monocropping—than autumn olive ever can or will.

A field is not a right or a wrong; it is a happening, a moment in time. The pathways we cut through the densest bramble of thicket are always narrowing, navigating the edge of chaos and control. And yet, we need the openings as much as we need the forests; a field is not a right or a wrong. A vast pre-colonial history of Indigenous human-forest relation-ships in Eastern North America included the knowledge that a clearing—a gap in the tree canopy—is an ethereal, shifting state, as alive as the creatures who move through it. Grass-land prairies—particularly the oak savanna thought to have evolved in collaboration with megafauna, man, and millennia—are some of the most productive ecosystems on the planet, creating food, fuel and fiber for wildlife and humans. Indigenous agricultural practice involved large-scale controlled burns to create clearings where growth of grasses, annuals, and forbs flourished. The new space and forage facilitated the movement of both large and

↑ Taylor Hanigosky *Basket in Process*

small animals that interacted with plant and soil communities in highly specific and evolved relationships of mutual benefit. After a period of time, the openings were allowed to regrow into mature forest again. This differs greatly from the Western concepts of commodity cropping, lawns, and empty parking lots, where an opening in the forest is continuously held in a fixed state.[2]

When we remove plants from a space and neglect to steward it back toward a healthy path of regrowth, we leave the soil, the existing seed bank, and the ghosts of the former plant and microorganism relationships in a state of stressed competition to fill the space. An opening is a moment in time and the evolu-

tionary intelligence of plants draws them into the understanding that the space must regrow. So they grow. But lacking healthy soil and relationships to build upon, the system is vulnerable to invasion, be it by pest, disease, or other invasive species. Enter autumn olive to repair the soil, slow erosion, and feed wildlife. We engineer our own disasters.

Right in the center of the only modest clearing left on our nine-acre parcel is a thicket of autumn olive. We hope to build a garden in this rare sunny space. After a year of listening to the wind shake a curtain of silver leaves, we oiled a chainsaw to bring down this productive, ripening patch and replace it with another. The grief of our reactive circumstance blew out

Taylor Hanigosky *Shoots in water*

from the chainsaw blade in a stream of soft sawdust. In acknowledgement of complicated and tangled histories, we must step into our role as population control whether we like it or not. Our role was chosen for us, as was autumn olive's, and our destinies are linked. A field is not a right or a wrong.

Inside the architecture of the thicket, I can see where the parts of myself seduced into destruction grow. I can see the part of myself that cut down the last American chestnut tree. I can see the part of myself that shot down the last bison. I can see the part of myself that dammed the river and planted the first autumn olive. The mistakes of my ancestors are mine. The motion of the ax and the plow are encoded in my body's memory. I have come here—to the field—to surrender. I have come here to eat the autumn olive berries and drink a tea of her leaves.

Our histories are not buried beneath inanimate time and earth. These legacies are alive and composting amidst an ecosystem of riotous, molecular life. In the wintery clarity of recognition, we can see the shape of the land and reconnect the pathways. We can rebuild the relationships and relate them, with autumn olive, so she may step into a new role of her choosing. We can rebuild the relationships between ourselves and this ecosystem so we can leave the role of destroyer behind for something new. ○

Notes

1. Kathy Smith and Annemarie Smith. "Controlling Non-Native Invasive Plants in Ohio's Forests: Autumn Olive and Russian Olive" (Columbus: Ohio State University Extension, 2012).
2. Eric Toensmeier and Dave Jacke, *Edible Forest Gardens*, vol. 1 (White River Junction: Chelsea Green, 2008).

Tatiana Dolgushina

WE HAD KNOWN LOVE

dedushka showed love
by picking the almonds
off the trees in Ukraine
in the countryside
and filling a potato bag
until it was full
with his love, imagine
a potato bag spilling
with almonds sitting in
our apartment in the city
my friend and i were 7
when we squatted on the
apartment floor
with a hammer in our hands
and hammered them open
right on the floor
for all the neighbors to hear
"this is amazing" my friend
kept saying, due to all the
abundance of seeds
for us to eat

she, a child of a crumbling
nation, was always amazed
at abundance
at having something for herself
as if she was stealing
and putting it in her mouth
instead of the gifts
that life brings to children
we were 7
and we knew already
a potato sack full of seeds
was something you hide
from the neighbor below
when he starts hitting
the heating pipes that
connected our apartments
due to our hammering noise
he didn't know
we were sitting on treasure
he didn't know
we were filling our cheeks
with dedushka's love
he didn't know
we had known love
already, with a hammer
in our hands,

he didn't know
we had seen the ruin
of everything around us
and chose to seek our
shelter, just like him
with seeds inside our cheeks
in case anybody comes
to take it all
away

Lavinia Currier

ENCOUNTERS WITH STAGS

Encounters with stags are frequent on Molokai. There was the stag my dog Kiawe chased and injured so cruelly that I had to cut his throat with a machete by moonlight; the stag so weak it chose to shelter from a storm under the eaves of my house (the dog killed him at dawn); the orphaned fawn that I carried home on my saddle and bottle-fed until it was big enough to eat grass who died of fright. But here is the story of a stag whose life we saved.

I was accompanied by a friend, F, and her son, V, whose birth I had attended thirty-two years ago, back home in a cabin that stood in a lush pasture of buttercups. Since they had arrived on island it had not stopped raining and we were feeling cooped up, so we decided to hike to the falls in the valley, although it was late and still drizzling. The swollen torrent obscured the ledge of rock from which it plunged, piercing the deep, dark pool with a loud, white splashing into which we leapt, swimming hard against the force of the current. Drying off, I found an apple in my backpack. No one had a pocket knife, but V produced a flint knife that he had knapped in an archeology class and quartered the apple with it. New threads of water appeared on the violently green slopes above us, witness to the days of deluge.

We walked back down the trail along the PVC pipe that parallels the ʻauwai—the old irrigation ditch—passing the brooding platform of black lava rock that was the valley's heiau, or temple, for ceremony and sacrifice in ancient times. Wanting to get to the beach road before dark, we took a shortcut through old stone walls now littered with rubbish left behind by drifters, hung with exotic vines, and void of understory due to the influx of axis deer. These deer arrived on Molokai in the mid-1880s as a gift from Asia to Hawaiʻi's King Kamehameha V, and now they overrun the island, denuding slopes that loosen dirt down the gulches in heavy rains, staining the ocean red.[1]

Standing in the trashy gloom, we noticed a magnificent stag, the largest I have ever seen, with wide, branching antlers and a copper coat dotted with white moons, resting on the ground. His legs were tucked under and his head high, watching us with wary eyes. The stag's antlers were ensnared in a tangle of thick vines, which wound around them twenty or thirty times, immobilizing him upright. He let us close, and we got to work to release him. F and I lifted what vines we could off his antlers, V sawing with the flint through the rest. The velvet of his antlers was cool and moist to the touch; one had broken off and bled slightly. Dusk fell, and after an hour he was free. He lay his head flat on the ground, and we thought he too might die of shock, but his eyes were calmly studying us all the while. We sat him up, then helped him to his feet. He swayed, pitched forward to plunge into the vines again as if he had forgotten how to run or escape, and then, with the grace of a ballet dancer, he pivoted into the dark forest.

On our island, axis deer outnumber human inhabitants almost ten to one.[2] As an intro-

Alfred Edmund Brehm *Brehm's Life of Animals: The Axis Deer*
(image courtesy of American Museum of Natural History Library)

duced species, they are detrimental to the native Hawaiian forests. The forests are completely defenseless, never having had ungulates with which to evolve, lacking the thorns or toxins with which trees and plants use to defend themselves on the mainland.

Unlike other deer species, axis does are able to give birth year-round and may have three or four fawns each year. In times of extreme drought, which are now more frequent, emaciated deer stumble into yards on the already-dry west end of the island in search of sustenance

Jake Muise *Deer*

and water, only to collapse and die by the many hundreds.[3] Valleys normally verdant are now brown and devoid of understory, and when the rains do come, they wash the topsoil into the ocean, where it covers and suffocates the reefs.

We humans live by the generosity and sacrifice of others. This is nowhere more true than in the animal world, at whose expense our prosperity as a species has come. Wild mammals now only make up a mere 0.4 percent of the mass of all mammals, while domestic livestock predominate worldwide, and are largely treated in such a way that can be said to be nothing less than torture.[4]

And yet these deer are lovely animals. They are graceful, moving in large herds on the hillside above my house at night, a line of red eyes, barking to another herd on a far ridge. Their meat is tasty, and they supply a large part of Molokai's subsistence diet, on an island where 98 percent of all food is imported from the mainland.[5] Community hunts are devoted to reducing their numbers and supplying families with meat. On our ranch, hunters shoot forty or so deer each weekend, but still the herd grows. As a conservationist I know we need to harvest more deer, but as a student of Buddhism whose teacher saves the

life of each insect about to be trampled, slaughter does not sit well with me.

I wince at the volley of gunshots on the ridge at dusk, no consolation that I do not hunt myself, as I admire my son and my neighbors who do so expertly, while I also admire those—like my daughter—who have never tasted meat. On our ranch we farm biodynamically, using the preparations that Rudolf Steiner developed to address imbalances in a homeopathic way. We take the stags transparent bladder, considered a purifying organ, and stuff it full of wilted yarrow flowers, then hang it to dry in the sun. Having absorbed the light's energy, the bladder full of yarrow is buried in a clay pot to spend the winter in the garden, then dug up, diluted in water that has been stirred in a vortex, and sprayed on our orchards, crops, and pastures to potentiate the element of sulfur in our soil. Using such methods, animals are essential to biodynamic farming.

Considering the idea that all living beings— including, but not limited to, humans—should have a "natural right" to exist peacefully in their own habitats where they can find sustenance, I am interested in the various legal cases currently being brought to courts in the United States, the United Kingdom, and Europe arguing that members of the other-than-human world should have "personhood." These include cases on behalf of factory farmed animals, on behalf of a captive elephant in the Bronx Zoo, and even on behalf of natural places, such as a pond in Florida.[6,7] In the US, only "persons" are afforded rights—a designation corporations have been awarded, but animals, trees, and sacred places have not.[8] Those in opposition may do so not on the basis of its logic, but on the grounds that most humans do not yet enjoy those rights in practice, and that if animals—even plants and places—are given rights, human economic activity will be severely constrained.

We might wonder on Molokai that if axis deer were given rights, could we still harvest them and stop them from eating our forests? In traditional hunting societies, where animals were most often respected and considered persons, they were hunted and eaten, so I imagine the answer is yes, with respect.

But if non-human beings are not awarded their "natural right" of personhood, and our unchecked plunder of the earth continues, they will cease to exist and disappear from the earth as they already are in great numbers with increasing frequency, leaving our human descendants to inherit the haunted ruins of the natural world. The last Black-faced Honeycreeper, the diminutive poʻouli, followed thirty-six Hawaiian birds into extinction in November 2021.[9] Having had no viable habitat nor tree snails to eat in Maui's forests for many years, it quietly closed its remaining eye in solitary confinement. Gone forever from the islands, the poʻouli's flitting through the branches of the ōhiʻa trees on wing feathers of twenty subtle shades of brown, gone too, its song. They say a body loses twenty-one grams at death, the weight of the soul. I wonder, if we push wild animals past the brink to extinction, that mere 0.47 percent remaining gone, will the earth have lost its soul?[10]

My friends and I saved one stag, because it was the right thing to do, yet as a landowner I condone the killing of many deer, so that our forest can regenerate. As a Buddhist, is this justified? Probably not. Buddhist teachers have debated under which circumstances it could be morally correct to kill a murderous person to save others' lives, and it seems the threat must be immediate, obvious, unambiguous. But to the Hawaiian, Hāhai nō ka ua i ka ululāʻau—the rain follows the forest—so ecologically, the harvest of axis deer unambiguously lets the forest recover, which ensures the watershed, and thereby sustains all life.[11] ○

Notes

1. Marcel Honore, "Molokai's Fabled Axis Deer Are Starving to Death in Droves", *Honolulu Civil Beat*, January 25, 2021.

2. Honore.

3. Honore.

4. Hannah Ritchie and Max Roser, "Biodiversity," *OurWorldinData* (2021).

5. Catherine Cluett Pactrol, "Feeding Molokai Sustainably," *Molokai Dispatch*, June 3, 2014.

6. Lawrence Wright, "The Elephant in the Courtroom," *The New Yorker*, May 7, 2022.

7. Elizabeth Kolbert, "A Lake in Florida Suing to Protect Itself," *The New Yorker*, April 18, 2022.

8. Christopher Stone, "Should Trees Have Standing?—Toward Legal Rights for Natural Objects," *Southern California Law Review* 45 (1972): 450–501), 1972.

9. Helen Sullivan, "Extinction Obituary: Why Experts Weep for the Quiet and Beautiful Po'ouli," *The Guardian*, May 4, 2022.

10. Visual Capitalist, "Visualizing the Total Biomass of Every Animal on Earth," *The World Economic Forum* (August 2021).

11. Mary Kawena Pukui, "Olelo No'Eau: Hawaiian Proverbs and Poetical Sayings" (Honolulu: Bishop Museum Press, 1983).

Heather White

ROOTING IN A CHANGING CLIMATE

When the fire came this time, I threw open the gates to let our chickens out. Sixty birds do not transport easily in the best of times, let alone when you're given an hour to decide what of your life you want to save from wildfire. We gave them extra food, water, tearful, loving words, and an open pathway to their best chance at escape, should it come to that. The thought of the birds—small little creatures who I raised from infancy, left stressed and unknowing as we fled a nightmarish column of smoke—broke something inside me.

Living and farming in the Southwest carries an inherent risk of wildfire. I grew up learning about fire safety, forest health through con-trolled burns, chronic and increasing drought. I've tried to avoid the truth that precautions and prescriptions are not enough anymore. But we live in a matchstick forest, naming scars on the landscape as easily as any other land-mark: Radio, Pumpkin, Rodeo-Chediski, Schultz, Slide, Boundary, Museum, and, now, Tunnel.[1] We fled this fire pretty certain that neither the house nor the farm would survive. We'd gotten lucky too many times already. That night, on a friend's air mattress in a spare room, my throat aching from smoke and shouting to be heard over the winds, I cried for those birds.

• •

Farming is, by nature, entering into deep relationship with the land as you work together to cultivate nourishing food. Or, perhaps, that is the ideal that some farmers in particular

work toward, as proponents of industrial agriculture might have a different idea of what that relationship should look like. But small-scale sustainable farming requires rootedness and deep and consistent attention to place so that a fruitful relationship can blossom that ideally benefits farmers, land, and community. Coming into relationship with land in this way allows a deeper understanding of place, which extends into the ways one interacts with the world at large. For me, it has led to deeper forms of engagement, and of a connection to the natural world that I might not otherwise have had. A sacredness and an intimacy.

In early 2021, I shared conversations with several other young, white farmers in Flagstaff, Arizona, as a part of a larger project discussing whiteness, senses of belonging, and relation-ship to land. We discussed what it might mean to belong to a place, and if, as white settlers living in a settler-colonial society, we can and should seek belonging on stolen land.

Whiteness leaves many in its wake search-ing for belonging. The West has disrupted and disconnected people from traditional land bases, white people included—places that carry with them culture, traditional foods, and obligations of reciprocal relationship. We discussed the urgency for white people to reinvigorate those long-lost relationships with land, and what those relationships might look like as we move through occupied land in an era of climate disaster.

Together we laughed as we imagined new (old) ways of being with the world through

tending land, and the joy that being witnessed by a landscape can bring. Despite their desires, none of the white farmers I spoke with really felt like they belonged anywhere. Framing this

was a deep awareness of the uncertain future awaiting us all with climate change. Their deep fear of what is happening and what is here. This influenced their notions of rootedness,

and what it means to belong somewhere. They thought it was almost necessary to not be deeply rooted anywhere. Yes, belonging involves commitment to a place, a community, and carries an intention to stay and dedicate oneself to something. However, the way they spoke alluded to farming and land work as a gateway to belonging at large, and not specifically to any one place. A heaviness set into their shoulders, their gazes held a faraway look, and their speech slowed, holding longer pauses. Uprootedness, the need to be mobile, loomed in their thoughts as they were making decisions. The weight of what farming might mean in an uncertain future, wherein the futurity of this settler-colonial society is finally being called into question, was clear.

White people, especially landholders and workers, grapple with knowing we aid in upholding colonial systems of domination that facilitate the exploitation and destruction of the world. Four centuries ago, civilizations subjected to the perils of "discovery" and "improvement" lost their land and were forced to adapt to survive or suffer oblivion. Climate change now graciously provides colonizers and settlers our own sense of this quandary as natural disaster, encroaching climate refugees, and uncertain politics strip away our constructed notions of safety and security. We are losing our sense of home, and many will have to take flight to find somewhere else to land. White people trying to engage in alternative agriculture for the betterment of our communities must apply self-awareness to our positions to ensure we are not perpetuating the systems of domination we claim to renounce. Should we even want to belong on occupied Indigenous land? Where do white settler-farmers fit in a world ravaged by a rapidly changing climate? Where do we go when the land we try to belong

to is made hostile? *(I let the chickens out and ran away from the flames.)*

My dear friend, a Diné woman, is also reckoning with her own versions of these questions: What happens when it seems like She (the land) does not want us here anymore?

We are staring down the barrel of climate uncertainty the likes of which have never been witnessed by humanity. We do not know what our world is going to look like. The only certainty we have is tremendous upheaval, and how can you plan to spend the rest of your life in place when you do not even know if that place will exist in the next fifty years? This unmooring is not limited only to those young farmers in Flagstaff. Everyone is losing their sense of place in one way or another, whether due to changing politics, changing climate, gentrification, or the like.[2] Farmers are trying to figure out how to grow in this climate crisis, understanding that land tenure anywhere is not necessarily a guarantee. We are wrestling not only with the very practical limitations of water security or variability in a season's length, but also the existential uncertainty of our ability to live—let alone farm—anywhere. *(I unlocked the gates under a rain of ash and smoke.)*

What then does tending mean, if you are uncertain about the longevity of your existence in a place? How much labor and dedication do you give to a project you are likely to leave? Can an ethos of mobility be built into this new era of farming, juxtaposing necessary impermanence with the very permanent work of stewarding land? *(I gave the birds their best chance and then ran myself.)*

We talked about climate change being the end of the world, or at least the end of a world. For many people of color in the colonial present, the world has already ended. Mississauga Nishnaabeg writer Leanne Betasamosake Simpson describes through her Indigenous ontology that we are living in perhaps the Fourth World, and that there are many that have come and gone already.[3] The world ended when Indigenous sovereignty was systematically wiped from the face of this continent, replaced by a domineering culture of endless growth and consumption. The world ended when other continents were pillaged and their peoples enslaved, taken from their traditional lands. It has always been the end of the world for someone. It ends again now, with the collapse of our ecosystems and the inevitable collapse of the culture that caused it.

Perhaps where these young farmers are seeking to belong is not in a place, but in a new ethos of farming and working with land, looking to plant themselves in a wider system of relations, even as those systems are changing beyond recognition. White settlers experiencing the end of the world must unsettle ourselves and enter into relationship with the places in which we land, along with other beings that find home there as well. Often

the reason we have the privilege to move is because we have little obligation to the land on which we reside—we can move about freely and can expect to land wherever we deem appropriate, regardless of the impact on our would-be neighbors. White colonial society has claimed this land as its property; as a given right, already belonging to us. In this, perhaps the first step in our search for belonging is understanding that maybe we do not belong. Our presence depends on the continued displacement and repression of Indigenous people, continuing those cycles of violence. But to unsettle whiteness and reestablish generative relationship with the living world, we have to believe we can belong somewhere. We have to reject the settler expectation of belonging, and instead focus on what it might mean to belong. Our shallow-rootedness, an intentional choice to be mobile, might be cultivated with the intention of healing, of creating reciprocal relationships while we can, where we can, until we cannot any longer—whether we be pushed out, out-lived, or called elsewhere.

•••

Almost a week after the evacuation, I stood on our small farm to survey the damage. The house stood intact behind me. The hoop houses stood with minor damage in front of me. The chickens knew nothing of what happened and had stayed on the property the whole time—our sweet rooster saw to his girls and received a fat watermelon for his good work. The seedlings, left uncovered, did not survive, as the temperatures dropped to the low twenties while the evacuation order held. It almost felt like an insult, but if frost and wind damage were our only losses, we were blessed. Our luck, it seems, has not quite run out. The

fire stopped just short of our plot of land, the line holding about two hundred feet away. Two hundred feet of dirt and dried grasses stood between us and destruction. Many of our neighbors were not as lucky. I know we likely will not be again, with population increases straining this land even further, rain and snow becoming scarcer, and temperatures creeping higher and higher. And fire season is only getting longer. Six weeks later we would evacuate again under eerily similar circumstances for the Pipeline and Haywire fires.

So we're making plans to leave. In many ways these fires felt like an eviction notice, or perhaps, more kindly, this land pushing us out of the way of a speeding truck. We know we cannot stay here. There is incredible grief in that—I am mourning leaving my home and that this place I love so deeply is dying. Changing beyond recognition as another casualty of climate change. I think of my conversations with those young farmers as we make our plans.

In that, we are trying to understand a sense of impermanence in what can be considered very permanent work and finding that relationship can transcend it. Perhaps land work and relationships with place in the Anthropocene are meant to be moveable, while an ethos of otherwise is carried into the uncertain future. ○

Notes
1. Names of prominent wildfires in Northern Arizona from 1977 to 2022.
2. I think of Bruno Latour's *Down to Earth: Politics in the New Climatic Regime*, and his ideas about climate change being a "wicked universality." Climate change is making the world unrecognizable for *everyone*, albeit at different paces dictated by wealth and social insulation.
3. Naomi Klein, "Dancing the World into Being: A Conversation with Idle No More's Leanne Simpson." *Yes!*, March 6, 2013.

Timothy Furstnau

PACE RANAE

Adjusting the Culture of Speed

The first few times I mowed the grass around the ponds where I live, the mower cut more than just grass. I would feel a little thud as I went, and stop the machine to discover the body of a large frog underneath, shredded to pieces. It'd been sitting near the edge of the pond, as frogs do, and didn't jump out of the way. Once, I could consider it a fluke. But after it happened twice, I started thinking about what I might do to keep it from happening again.

Slow down. Adjust down the speed setting of the self-propelled mower so frogs might have more time to jump out of the way. *Be careful.* Inspect more closely the grass ahead of the mower, to see the frogs in time to stop and relocate them. *Warn them.* Walk around the pond first and try to flush out any frogs before I come through with a giant spinning blade. Or all of the above. But all of the above doesn't change much about the situation. They take the tool at hand, the lawnmower, for granted. They don't see past what is literally right in front of me.

Other responses involve reconsidering the tool itself. *Make pasture.* Get goats or sheep or other ruminants to graze the grass instead of cutting it. *Go manual.* Cut the grass with a nonelectric push mower or scythe. Plugging

in doesn't wash my hands of the dirty energy I get from the grid, so these options are appealing.

But all these most obvious adjustments share a problem: they take time. Like so many people, in order to stay afloat economically I can't currently care for livestock, manually mow a few acres of grass, or slow down the process in other ways. The more I thought about it, that first and most obvious response— the speed adjustment on my lawnmower— seemed an apt metaphor for a larger, more fundamental problem of time: a mismatch of speeds.

To elevate the problem to the level of a concept and coin a term, I'd like to propose *pace ranae* (pronounced "pah-chay rah-nay"), a suitably pretentious yet catchy term from the Latin for "speed of frogs." Frogs are already considered an indicator species for so many environmental issues, and environmental discourse has a tradition of using frogs metaphorically to refer to the speed of cultural change via the spurious "boiling frog" parable.[1] So frogs may be a worthy namesake for a mismatch of speeds in other contexts and scales.

What I mean by pace ranae might best be compared to the Marxian concept of a "metabolic rift," or differential processing of energy and material flows between class society and non-human ecologies. As one cause or effect of it, or one variant to do specifically with the movement of culture, and the ways time is constrained by economic pressures. In this seemingly trivial case, the pace of frog life seems out of sync with my lawn mowing.

Frogs have their own options, depending on different times or seasons, or perhaps according to species. When danger approaches, a frog's general strategy seems to be to remain in

place, stay as still as possible, and hope to go unnoticed. Conveniently, this is also their strategy for hunting, or being a danger in their own right to other creatures in the food web. Often they will stay motionless in their spot until you get surprisingly close to them, almost touching, before they retreat by jumping into the water.

By not questioning the lawn itself, or my desire to maintain it, I was adopting a strategy like the frog's, remaining motionless in the face of danger, even as it looms closer and closer. When I came into this land, I inherited the problem of maintaining it. And growing up in the culture I did, I inherited an expectation to see well-manicured lawns around houses. Agriculture, after all, from Latin *ager* 'field' and *cultura* 'growing, cultivation,' is a kind of culture applied to land. The problem of pace ranae is as much one of culture as it is of the material conditions and technologies of grass-cutting. The problem is not the speed or schedule, but the fact of my mowing; not how or when, but that I *mow*.

Pace ranae then becomes my name for the insufficiency, privilege, and danger of merely slowing down. The problem is not speed, but asynchrony between cultures. Not just the speed of culture, but the culture of speed.

The frogs, too, have a kind of inherited cultural norm. Since this pond is spring-fed, it is likely that some small body of water has been here for many years. Frogs are known to return to their place of birth to breed, using a variety of visual, auditory, olfactory, and even celestial and magnetic cues. For them it's habitual, if not instinctive. There is a community legacy here that far predates my own or other human use.

What I decided to do, in the end, is *something else*. A mix of things, actually. For my yard, I let the grasses grow around the edge of the pond, broadening the natural "ecotone," or border habitat, between water and terrestrial environments where diverse species find cover.

Germain Seed and Plant Company *Lawn Mowers Our Special* (image courtesy of US Department of Agriculture, National Agricultural Library)

I let some lawn remain—for the time being for use in my composting, and for walking around the pond with less worry about ticks. But I'm slowly converting other grassy sections of the yard to pollinator gardens and other planting areas. With gradual adjustments, my vision is becoming less fixed to the conventions of "landscape," and my presence is gradually more in step with the pace of the others that are here.

I also started volunteering with New York's Amphibian Migrations and Road Crossings Project, helping frogs, salamanders, and other amphibians cross roads during the critical spring breeding season. What sounds like a cliche of do-goodism is actually an important service for conservation science, as participants also collect data on thousands of specimens, both live and dead. For dozens of species, it's a real-life version of the classic video game *Frogger*, demonstrating quite plainly the difference in speeds between amphibians and automobiles. I like to think of it as a seasonal ritual, an intentional cultural custom to replace the many hollow holidays I inherit pre-inscribed into the calendar year. Together with hundreds of others across the state, we wait for the perfect warm, rainy conditions known as Big Night, when countless creatures seize the moment to make their annual journey to ephemeral vernal pools. In an absurd choreography of safety vests and headlamps, we attempt to count them. And, yes, sometimes help them cross the road. And in the process—sensing the weather, learning their routes—we become more attuned to place.

And in the community, beyond my yard, I also helped start a tool lending library called Toolshed Exchange. Unlike many tool libraries, our inventory does not include a lawnmower. Our decision not to stock one was intentional, as both a practical appeal to apartment-dwellers and a subtle cultural and infrastructural prod, encouraging our members to reconsider their land use, speed, and culture. There's the old saying that with a hammer in your hand, everything looks like a nail. Without a lawnmower in our local tool library, maybe everything will start to look less like a lawn.

The tool library is part of a larger project called Toolshed, which functions in part as a kind of library for useful ideas. An idea is also a kind of tool. Orthodox economists like Milton Friedman are wrong about a lot of things, but he was right, about one thing: "When [a] crisis occurs the actions that are taken depend on the ideas that are lying around. That, I believe, is our basic function: to develop alternatives to existing policies, to keep them alive and available until the politically impossible becomes the politically inevitable."[2] In what is increasingly being referred to as a climate crisis, Toolshed Exchange is an idea of a real sharing economy, of commoning and re-commoming, that we're trying to keep lying around, ready to be picked up and used. ○

Notes

1. Arising from 19th century experiments purporting to show that a frog put in slowly boiling water will not jump out before being cooked alive, the "boiling frog" is often used as a metaphor for awareness of the negative consequence of gradual, imperceptible change. Though the point is well taken, modern scientists have called into question the original studies.
2. Milton Friedman, *Capitalism and Freedom* (Chicago: University of Chicago Press, 1982), xiv.

Alli Maloney *Under drones at the southern border, Kumeyaay/Paipai (Akwa'ala) Land*

IMAGE CREDITS

This is a noncommmercial, nonprofit publication. To reprint original work, request permission of the contributor.

3 Fallen Fruit (David Burns and Austin Young), *Fruits from Garden and Field Detail* (image courtesy of the Victoria and Albert Museum London), 2019, Digital print

4 Anonymous, *Ex libris van Aloise Kučíka (1900–1999)*, Rijksmuseum.

9 Unknown photographer, *Mississippi Flood*, 1929, Black and white photograph. (image courtesy of Library of Congress)

11 Renée Rhodes, *Foxtail and Fescue*, 2021, Digital photograph

12 Alli Maloney, *Along the Ohio, Adena/Hopewell/ Shawandasse Tula (Shawanwaki/Shawnee) Land*, 2020, Photograph

15 Alli Maloney, *Stoned with La Doña at Blue Mountain Center, Haudenosaunee/Kanien 'kehá: ka Land*, 2019, Photograph

16 Colin Sullivan-Stevens, *Grey Lodge*, 2021, Pencil drawing

18 Colin Sullivan-Stevens, *Reversing Hall*, 2021, Pencil drawing

21 Gavin Zeitz, *Map of Smithereen Farm*, 2021, Digital drawing

22 Ginny Maki, *Pembroke Town Map*, 2021, Digital Map

24 Alli Maloney, *Smithereen Farm, Passamaquoddy Land*, 2019, Photograph

25, 29 Adam and Charles Black, *Sidney Hall and William Hughes The Solar System and Theory of the Seasons (excerpt) from General Atlas Of The World: Containing Upwards Of Seventy Maps*

33 Alivia Moore, *My daughter harvesting cranberries*, 2022, Digital photograph

34 June Sapie, *Medicines in the birchbark bowl*, 2022, Digital photograph

37 Hilary Irons, *White Window*, 2015, Oil on panel

38 Brett Ciccotelli, *Dennys River Restoration*, 2021, Digital photograph

39 Brett Ciccotelli, *Alewives*, 2018, Digital photograph

40 Brett Ciccotelli, *Restored Dennys River at Meddybemps*, 2022, Digital photograph

41 Maia Wikler, *Salmon Heart*, 2021, Digital photograph

42 Maia Wikler, *Salmon Drying*, 2021, Digital photograph

44 Alex Plowden, *MycoBUOYS* workshop at Smithereen Farm, 2022, Digital photograph

46 *Sorted particles of microplastic.* © FAO 2022. Everything you ever wanted to know about plankton and microplastics sampling

47 *Neuston Net.* Image courtesy of Islands in the Sea 2002, National Oceanic and Atmospheric Administration/ Ocean Exploration and Research

50 Lori Rotenberk, *Henry on the farm*, 2021, Digital photograph

52 Henry Brockman, *Soil*, 2021, Digital photograph

52 Henry Brockman, *Soil holding water*, 2021, Digital photograph

53 Terra Brockman, *Henry's ten-year journal*, 2021, Digital photograph

54 Lori Rotenberk, *Aozora Brockman*, 2021, Digital photograph

58 Amory Abbott, *Flood 3*, 2021, Charcoal and white charcoal pencil on grey Stonehenge paper

60 Amory Abbott, *Flood 1*, 2021, Charcoal and white charcoal pencil on grey Stonehenge paper

62 Unknown photographer, *Before and After the Hurricane of September 1938*, 1938. (image courtesy of Westport Historical Society)

63 Unknown photographer, *Site of same cottages (above) after the storm*, 1938. (image courtesy of Westport Historical Society)

64 Harry Bloomingdale, *Trafford and other familiar property before the storm*, 1938. (image courtesy of Westport Historical Society)

64 Harry Bloomingdale, *Site of same cottages (above) after the storm*, 1938. (image courtesy of Westport Historical Society)

64 Harry Bloomingdale, *East Beach Horseneck*, 1938. (image courtesy of Westport Historical Society)

64 Harry Bloomingdale, D*igging out the road on East Beach*, 1938. (image courtesy of Westport Historical Society)

65 Emily Vogler, *Zones of beach determine public access*, 2022, Digital illustration

67 Briana Waltman, *Bill Covering Parsnips for the Winter*, 2021, Watercolor

69 Cafe Ohlone, *Logo*, Digital Illustration

70 Cafe Ohlone, *Quail eggs*, 2022, Digital photograph

72 Cafe Ohlone, *Stew*, 2022, Digital photograph

75 Suzanne Husky, *Smokey the Beaver*, 2022, Watercolor

76 Hannah Althea, *Nature's Natural Builders: Beavers home at Errol Heights Wetlands in Portland, Oregon*, 2021, Digital photograph

76 Hannah Althea, *Nature's Natural Builders: Cob "Sanctuary" at Planet Repair Institute in Portland, Oregon*, 2021, Digital photograph

78 Mike Iocona and Scott Kessel, *Refuge Yurt*, 2022, Pencil and eraser on paper

83 Barbara Rose, *Farmhouse and lower section of the farm*, Digital photograph

83 Barbara Rose, *Sprouting ironwood and palo verde seeds with dry mesquite pods*, Digital photograph

84 Robert Dash, *Fava Cover Crop*, 2019, Archival print

157 Wezel (Mustela nivalis), *Anselmus Boëtius de Boodt*, 1596–1610. Public Domain.

160 Don Tipping, *Garlic*, 2016, Digital photograph

163 H.H. Iltis & Doebley, *Taxón: Zea mays subsp. mexicana* (image courtesy of The Trustees of Indiana University)

164 National Academy of Sciences, *Teosinte and Maize*. (image courtesy of PNAS: Proceedings of the National Academy of Sciences of the U.S.A.)

169 Supermrin, *Collected images from the Condition Reports, Oakland Cultural Heritage Survey, City Planning Department* (images courtesy of Oakland City Hall Archives), 1985, Digital college

171 Supermrin, *Frank Ogawa Plaza*, 2020, Digital collage using Google maps images

172 Jessica Fertonani Cooke, *Braiding Field performance, Governors Island, NY*, 2021, Performance documentation

174 Supermrin and Xenia Adjoubei, *Material Sample 1 with biomaterial, grass clippings, cow ribs, performance remains*, 2022

177 Supermrin, *FIELD (leaves)*, 2022, Biomaterial and turmeric

178, 179 Joel Catchlove, *On Weeds*, 2022, Ink on A3 Paper

181 Chad Westbrook Hinds, *Trousseau vineyard in western Siskiyou County*, 2022, Digital photograph

183 American Homes and Gardens, *Tomato vines trained against a wall*, 1912. (image courtesy of the Smithsonian Libraries)

184 Briana Waltman, *Perre in the Elderberry*, 2020–2021, Watercolor

189 Bradford Torrey, *Everyday Birds: Scarlet Tanager (i. Male, ii. Female)*, 1901. Image courtesy of The Library of Congress

192 Milo Vella, *Ladder and Young Peach*, 2020, Watercolor and pencil on paper

195 Milo Vella, *Thank You, Bo!*, 2020, Watercolor and pencil on paper

196 Obediah B. Stevens and Robert F. Wright, *Miller, Raspberry* (image courtesy of Georgia Department of Agriculture)

199 Dorothea Lange, *El Monte federal subsistence homesteads. Three-room house seventy dollars and seventy cents monthly. Rent to apply on purchase. Four in family. Father, carpenter, earns seventy dollars monthly. California* (image courtesy of the New York Public Library Digital Collections)

200 Briana Waltman, *Peeling Soldier Beans*, 2021, Watercolor

203 Omar de Kok-Mercado, *Yellow coneflower*, 2020, Digital photograph

204 Omar de Kok-Mercado, *An oak tree in the process of being released*, 2022, Digital photograph

205 Omar de Kok-Mercado, *Monoculture corn dominates the Iowan landscape*, 2021, Digital photograph

206, 207 Omar de Kok-Mercado, *A prairie strip next to corn*, 2021, Digital photograph

208 Omar de Kok-Mercado, *A reconstructed tallgrass prairie*, 2021, Digital photograph

210 George Shaw and Frederick P. Nodder, *The Garden Snail*, Helix hortensis, 1789 from *The Naturalist's Miscellany: or Coloured Figures of Natural Objects; Drawn and Described Immediately from Nature. Vol. 1.* (image courtesy of the Biodiversity Heritage Library)

212 Gina Rae La Cerva, *Threshing Rice in Borneo*, Digital photograph

213 Adrien Segal, *Wheat Mandala Series: Puccinia striiformis*, 2018, Digital Collage printed on acrylic

213 Gina Rae La Cerva, *Fish and Rice*, Digital photograph

215 Sonomi Obinata, *Sonomi with dandelion painting*, Digital photograph

216 Elizabeth Blackwell, *Dandelion* (image courtesy of The New York Public Library Digital Collections)

218 Hans Kern, *Escape from Scare City*, 2022, Ink on paper

219 Adrien Segal, *Wheat Mandala Series: Man-Made Famine*, 2018, Digital Collage printed on acrylic

222 Rose Robinson, *The Chatterbox - Putnam Camp: Keene Valley, NY*, 2021, Acrylic painting

224 Samantha Winship, *Preparing for beekeeping*, 2019, Digital photograph

226 Briana Waltman, *Raking Blueberries*, 2020–2021, Watercolor

227 Nina Montenegro, *Harvesters*, 2019, Collage

228 Emily C-D, *Trueque Translocal de Semillas Libres (Translocal Free Seed Swap)*, 2021, Collage with watercolor on paper, bean and corn seeds

230 Linley Dixon, *Mother Trees: Becky Weed with her sheep guardian, Ruby, at Thirteen Mile Farm in Belgrade, Montana*, Digital photograph

231 Linley Dixon, *Mother Trees: Real Organic Project Co-Director Linley Dixon with Jessica McAleese at Swift River Farm in Salmon, Idaho*, Digital photograph

232 Kristin Leachman, *Bridalveil Fall (Yosemite, California)*, 2020, Gouache on wove paper. Previously published in *American Forests Magazine*, 2022

235 Rimona Eskayo, *Life and Death and In-Between*, 2020, Handcut paper

236 Chuck Monax, *New Website*, 2020, Digital illustration

237 Chuck Monax, *Self-Reliance*, 2022, Digital illustration

239 Melody Overstreet, *Waterfall*, 2019, Cyanotype print. Previously published in *The Freshwater Review*, 2022

240 Melody Overstreet, *Gorge*, 2019, Cyanotype print

244, 245 Filip Van Dingenen, *Algae Diplomacy - Almanac Series* (images courtesy of Waldburger Wouters Brussels and artist), 2021, Aquarelle on paper

247 Cielo Sand Hodson, *Bounty from the Mother Earth*, Digital photograph

248 Phil Ross, *Splash*, 2009, Drawing

253 Nanea Lum, *Loli i ka ʻūmalu*, 2022, Charcoal and oil on canvas

→ Alli Maloney
Resist means to grasp from the root, Canarsie/Munsee Lenape Land

INDEX OF CONTRIBUTORS

Madelaine Corbin
Artist, educator
Detroit, Michigan
madelainecorbin.com + @madelainecorbin

Phil Cordelli
Farmer
Lakewood, Colorado
commonnamefarm.org

Lavinia Currier
Wildlife advocate, filmmaker, writer
Molokai, Hawai'i
puuohoku.com

Robert Dash
Photographer, educator
Deer Harbor, Washington
robertdashphotography.com

Omar de Kok-Mercado
Soil scientist, regenerative grazier
Pilot Mound, Iowa
@eolian.mollisol

Felipe Delfino
Artist
Brattleboro, Vermont
felipedelfino.com + @felipefdelfino

Linley Dixon
Real Organic Project co-director
Durango, Colorado
realorganicproject.org

Tatiana Dolgushina
Writer, science teacher
Springfield, Massachusetts

Lynnell Edwards
Poet
Louisville, Kentucky
lynnell.edwards.com + @lynnelledwards

Rachel Edwards
Chinese medicine, and Taiji Quan practitioner
Lincoln, Vermont
blackmoonharbor.com

Nina Elder
Artist, researcher
Datil, New Mexico
ninaelder.com

Ingrid Ellison
Artist
Camden, Maine
ingridellison.com + @paintinginthebarn

Rimona Eskayo
Artist, optimistic agitator
Portland, Oregon (Chinook land)
rimoskyo.com

The Farwoods (Nina Montenegro and Sonya Montenegro)
Artists
Bloomington, Indiana
thefarwoods.com + @thefarwoods

Jacob W. Forquer
Poet
Columbus, Ohio

Timothy Furstnau
Writer, curator
Windham, New York
timothyfurstnau.com

Amy Franceschini, Futurefarmers
Artist
San Francisco, California + Gent, Belgium
futurefarmers.com

Emily Gaetano
Textile artist, horticulturist
Maastricht, The Netherlands

Tanja Geis
Artist, graphic designer
Oakland, California
tanjageis.com

Toni Gentilli
Artist, anthropologist, naturalist
Albuquerque, New Mexico
tonigentilli.com + @phytomorphologie

Tom Giessel
Farmer, amateur historian,
member of National Farmers Union
Larned, Kansas

Zohar Gitlis
Former farmer, student
New York, New York
@zoharzip

Madeleine Granath
Farmer
Albuquerque, New Mexico

Christine Hadsel
Curtains Without Borders director
Burlington, Vermont
curtainswithoutborders.com

Taylor Hanigosky
Land steward, thicket dweller,
Wild Altar Farmstead partner
Stuarts Draft, Virginia
wildaltarfarmstead.square.site +
@wildaltarfarmstead

H.e. Haugenes
Land steward, beekeeper, multidisciplinary artist
Cold Spring, New York
@waterb.ug

Christine Heinrichs
Writer
Cambria, California

Elizabeth Henderson
Farmer
Rochester, New York
thepryingmantis.wordpress.com

Olivier Herlin
Ecologist, mycologist
Toronto, Ontario
@collectiveforager

Christine Hill
Artist
Vermont
nofavt.org + @nofavermont

Suzanne Husky
Artist, beaver activist
San Francisco, California
suzannehusky.com

Leke Hutchins
PhD candidate
Honolulu, Hawai'i
@indigenizesci

Mike Iacona
Guilford, Vermont
dryurts.com

anna ialeggio
Artist, educator
Ithaca, New York
aialeggio.net

Oliver Kellhammer
Artist, writer, and researcher
New York City, New York
www.oliverk.org

Jennifer Monson
Choreographer, improvisor
New York City, New York
Ilandart.org

Hilary Irons
Gallery & exhibitions director
Portland, Maine
@h.irons.h

Kim Kelly
Independent labor reporter and author
Philadelphia, Pennsylvania
@grimkim + @kimkellywriter

Scott Kessel
Artist
Middleton, Connecticut
@kesselarts

Hans Kern
Illustrator, Citizen's Assembly advocate
London, United Kingdom
hanskern.earth

Nance Klehm
Soil scientist, ecologist
Chicago + Drifless Region, Illinois
socialecologies.net

Gina Rae La Cerva
Geographer, environmental anthropologist,
author
Santa Fe, New Mexico
ginaraelc.com + @feastingwild

Kristin Leachman
Artist
Pasadena, California
kristinleachman.com + @kristinleachman +
@fiftyforests

Shauna Lee Lange
Artist
Jolon, California
steamcreatives.com + @steamcreatives

Falipa Lilias
Farmer
Point Reyes Station, California

Rose Linke
Writer, editor
San Francisco, California
roselinke.com + @theroselinke

Nanea Lum
Artist, master in painting
Honolulu, Hawai'i
nanealumpaintings.com

Teddy Macker
Carpinteria, California

Maia Wikler
PhD candidate, film director, writer
Vancouver Island, British Columbia
@maiareillyw

Austin Miles
Environmental scientist
Boston, Massachusetts

Chuck Monax
Farmer, illustrator
Madison, Wisconsin
@farmpunktoons

Alivia Moore
Panawáhpskewi, a person of the river where the
rocks widen
So-called Northport, Maine

Leaf Myczack
Farmer, teacher
Pilot, Virginia
sustainability-teaching-farm.com

Veronica Nehasil
Research assistant
Livonia, Michigan
@veronica.cheyenne

Mary O'Brien
Writer, artist
Fairfax, California
watershedsculpture.com

Casey O'Neill
Farmer
Mendocino County, California
happydayfarmscsa.com + @happydayfarms

Sonomi Obinata
Biodynamic farming practitioner
Hudson + Southold, New York
@sonominata

Sharifa Oppenheimer
Author, teacher
Charlottesville, Virginia
sharifaoppenheimer.org + @sharifaoppenheimer
+ @litanyofwildgraces

Melody Joy Overstreet
Artist, writer, printer, weaver, community educator
Unceded Awaswas territory in so-called Santa
Cruz, California
reciprocalfield.com + @melodyoverstreet

Luz Paczka
Ecologist, health designer, researcher
Toronto, Canada
collectiveforager.ca + @luzpaczka

Carol Padberg
Artist, educator, Iron and Indigo Sheep Farm,
Ranchos de Taos, New Mexico
carolpadberg.com + @nookfarmhouse

Eva Parr
Farmer
Point Reyes, California
@common_loon

Poki Piottin
Farmer
Dilia, New Mexico
milabrazos.org

Benjamin Prostine
Writer, farm worker
Soldiers Grove, Wisconsin

Brenna Quinlan
Permaculture illustrator
Denmark, Western Australia
brennaquinlan.com + @brenna_quinlan

Rose Robinson
Student
Middlebury, Vermont
thelifeofmugs.com + @thelifeofmugs

Melina Roise
Farmer, curator
Hudson Valley, New York
@melina.roise

Barbara Rose
Student of the Sonoran Desert
Pima County, Baja Arizona
beantreefarm.com

phil ross
Chief technology officer
Bay Area, California

Lori Rotenberk
Journalist
Chicago, Illinois
lorirotenberk.net

Michele Scott
LWOP advocate
Fallbrook, California
@micheleInvictus

Adrien Segal
Artist
Oakland, California
adriensegal.com

Maria Sgromo
Independent art filmmaker, beginning shepherd
Oregon
@maria.sgromo

Abdullah Shihipar
Writer, public health scholar
Providence, Rhode Island
@AShihipar

Dimitra Skandali
Multimedia artist
Paros, Greece
dimitraskandali.com

Gary Snyder
Poet
Nevada City, California

Mariee Siou
Singer, songwriter
Nevada City, California
marieesiou.com + @marieesiou

Kacey Stewart
Writer, educator
Chestertown, Maryland
@puc_puggy

Sharon Stewart
Cultural landscape photographer
Chacon, New Mexico
sharonstewartphotography.net

Aubrey Streit Krug
Writer, teacher, agricultural researcher
Smoky Hills Region, Kansas
landinstitute.org

Supermrin
Artist, educator
Cincinnati, Ohio
streetlight.space + @supermrin

Sandra Taggart
Artist
Brooklyn, New York
sandrataggart.com + @sandrataggart_art

Don Tipping
Seed keeper, farmer
Williams, Oregon
siskiyouseeds.com

Liz Toohey-Wiese
Artist, professor
Vancouver, British Columbia
liztoohey-wiese.com

Samantha Winship
Farmer, beekeeper
Winston Salem, North Carolina
mothersfinesturbanfarm.com + @mothersfines-turbanfarm

Christopher Winslow, PhD
Director at Ohio Sea Grant and Stone Laboratory,
School of Environment and Natural Resources
Columbus, Ohio
ohioseagrant.osu.edu

Filip Van Dingenen
Brussels, Belgium
fantaman.net

Sue Van Hook
Mycologist, educator, healer
Cambridge, New York
suevanhook.com + @ssvanhook

Tony VanWinkle
Assistant professor Greensboro,
North Carolina

Milo Vella
Student
Payahuunadü (Owens Valley, California)
as.cornell.edu/milo-vella

Vincent Medina and Louis Trevino
Mak-'amham/Cafe Ohlone co-founders
Berkeley, California
makamham.com + @makamham

Emily Vogler
Associate professor
Providence, Rhode Island +
Albuquerque, New Mexico

Danielle Walczak
Farmer, cook, baker
Portland, Maine
daniellewalczak.com

Mo Walrath
Willow coffin + basket maker,
community death care + threshold worker, artist
Port Townsend, Washington
woventhresholds.com

Briana Waltman
Journeyperson
Belfast, Maine
nowandnear.blogspot.com

Vincent Waring
Artist, craftsman
Santa Cruz, California
vinwaring.com + @vinwaring

Linda Weintraub
Artist, curator, author, homesteader
Trumansburg, New York
lindaweintraub.com

Chad Westbrook Hinds
Winemaker
Fort Jones, CA
iruaiwine.com

Heather White
Farmer, artist
Peaks Region (Flagstaff), Arizona
refugiagardens.org

Morgan Whitehead
Artist, bookbinder, writer
Richmond, Virginia
morganwhitehead.net

Casey Whittier
Artist, educator, civic scientist
Kansas City, Missouri
@caseywhittier

Margaret Wiss
Choreographer
New York City, New York
margaretwiss.com + @wiss.co

Zach Withers
Farmer
San Antonito, New Mexico
polksfolly.com

Zev York
Student of trees and humans
Middlebury, Vermont

Ethan Young
Ranching apprentice
Beaumont, Kansas

Connie Zheng
Artist, writer, filmmaker, PhD student
xučyun / Oakland, California
conniezheng.com + @yconniezheng

Driessens & Verstappen
Artist couple
Amsterdam, The Netherlands
Driessensverstappen.nl

Jin Zhu
Artist, video-maker
Bay Area, California
jin.is + @killeryellow

SUBMISSIONS

Che Sarà Sarà.   H.T.Dunn.

We begin at the end and end at the beginning.

Was your story missing from *Vol. VI*? Curious what comes next? We are too! We wonder what our collective consciousness can feel, predict, and gather in our foreboding bodies. As the climate changes and shifts around us, the news reports tell us that we are too late, and the cascading crisis points of fire and flood tip the scales all around us, we wonder what wells up in your body, in the body of the land, and in actionable response.

A premonition is a strong feeling that something is about to happen—usually something quite unpleasant. What are you noticing and sensing? What other premonitions abound? Think futures, futurisms, reclamations of apocalyptic thought. What kinds of premonitions swirl in the mind and how do they materialize in the present of your farming, landwork, and shifting relations to place?

Tune into our website, blog, newsletter, and social media to catch the open call for submission to *Volume VII: Premonition.* Or share your visions and intuitions by email: almanac@greenhorns.org to be considered for our 2025 edition of *The New Farmer's Almanac*. We look forward to hearing from you!

ABOUT THE GREENHORNS

Alli Maloney   *Jone's strawflowers at the Kirkendall house*

The Greenhorns works to promote, recruit, and support the next generation of farmers through grassroots media production. Our role is to explore the context in which new farmers face the world, through publications, films, media, and events—and by promoting the important work being done by so many organizations, alliances, trusts, and individuals around the world.

Greenhorns is based in Downeast Maine along the Pennamaquan River in the old Pembroke Ironworks. Our campus is spread out around town with a carpentry shop, boat shop, mycological lab, agrarian library, and many living and art spaces. There's always something new getting going, and we welcome potential collaborators to come for a visit.

Stop by greenhorns.org to watch our EaRtHLIFE series, download a guidebook, register for a seaweed webinar, or order all five volumes of *The New Farmer's Almanac*. Join our mailing list for monthly news of naturalist trainings, EaRtHLIFE releases, the next *New Farmer's Almanac*, and invitations to adventures on land, sea, and internet.